# Power Generation and the Environment

# SCIENCE, TECHNOLOGY, AND SOCIETY SERIES

*(Formerly Monographs on Science, Technology, and Society)*

1 Eric Ashby and Mary Anderson *The politics of clean air*
2 Edward Pochin *Nuclear radiation: risks and benefits*
3 L. Rotherham *Research and innovation: a record of the Wolfson Technological Projects Scheme 1968–1981*; with a foreword and postscript by Lord Zuckerman
4 John Sheail *Pesticides and nature conservation 1950–1975*
5 Duncan Davies, Diana Bathurst, and Robin Bathurst *The telling image: The changing balance between pictures and words in a technological age*
6 L.E.J. Roberts, P.S. Liss, and P.A.H. Saunders *Power generation and the environment*

# Power Generation and the Environment

## L.E.J. Roberts

*School of Environmental Sciences,*
*University of East Anglia, Norwich*

## P.S. Liss

*School of Environmental Sciences,*
*University of East Anglia, Norwich*

and

## P.A.H. Saunders

*Environmental and Medical Sciences Division,*
*Harwell Laboratory,*
*UKAEA, Oxfordshire*

OXFORD   NEW YORK   TOKYO
OXFORD   UNIVERSITY PRESS
1990

Oxford University Press, Walton Street, Oxford OX2 6DP
Oxford New York Toronto
Delhi Bombay Calcutta Madras Karachi
Petaling Jaya Singapore Hong Kong Tokyo
Nairobi Dar es Salaam Cape Town
Melbourne Auckland
and associated companies in
Berlin Ibadan

Oxford is a trade mark of Oxford University Press

Published in the United States
by Oxford University Press, New York

British Library Cataloguing in Publication Data
Roberts, L. E. J. (Lewis Edward John), 1922–
Power generation and the environment.
1. Energy supply. environmental aspects
I. Title.   II. Liss, Peter S. (Peter Simon)   III. Saunders,
P. A. H.
333.7911
ISBN 0–19–858338–9

Library of Congress Cataloging in Publication Data
Roberts, L. E. J. (Lewis Edward John), 1922–
Power generation and the environment / L.E.J. Roberts, P.S. Liss,
and P.A.H. Saunders.
(Science, technology, and society series)
Includes bibliographical references. (p.   ).
1. Electric power production—Environmental aspects.   2. Electric
power production—Environmental aspects—Great Britain.   I. Liss,
P.S.   II. Saunders, P. A. H.   III. Title.   IV. Series: Science,
technology, and society series (Oxford, England)
TD195.E4R64   1990
333.79'3214—dc20   90–6859
ISBN 0–19–858338–9

Set by Hope Services (Abingdon) Ltd
Printed in Great Britain by Bookcraft Ltd,
Midsomer Norton, Avon

# Preface

Energy policy is becoming one of the central questions on the political agenda. The logistical and economic problems involved in satisfying the energy needs of an expanding world population and of closing the gap between the energy consumption of the richest, industrialized nations and the poorest, agricultural nations have been compounded by the realization that increasing use of fossil fuels may cause unacceptable levels of global environmental damage and, possibly, irreversible climatic change.

These considerations are bound to impinge on the technologies used to generate electricity. The consumption of electricity is likely to grow because of its versatility, cleanliness, and efficiency at the point of use. In this book we have attempted to summarize the most important local and global environmental effects of the large-scale generation of electricity. After a historical chapter which traces the reasons for the pattern of generation we have today, we deal with generation from fossil fuels, from nuclear energy, and from some of the more promising 'renewable' technologies.

In many cases, the estimation of environmental effects cannot be made in a rigorous, quantitative way. For example, attempts to estimate the extent of global climatic warming due to increasing concentrations of carbon dioxide, in particular the local effects of such a change, are subject to considerable uncertainty because of the complexities of the world's atmospheric system and its interactions with the oceans, land, and ice surfaces. Again, any attempt to calculate the environmental risks arising from nuclear power depends on an estimate of the risks associated with low doses of radiation, which cannot be measured by direct experiment and will remain a matter of controversy. It is, indeed, very often the case that decisions on environmental matters have to be taken on uncertain evidence. We have therefore included a considerable amount of technical detail, although the book is aimed at the general reader rather than the technical expert, in order to illustrate as fairly as possible the present state of knowledge in the various subjects. The examples given of current practice and of regulatory rules and procedures mostly refer to the United Kingdom, but are meant to illustrate questions which are of world-wide significance.

We do not attempt to reach optimal conclusions, but rather to set out the evidence in sufficient detail for readers to reach their own conclusions on the weight that should be ascribed to environmental considerations, as against questions of capital cost and security of supply, when decisions on new plant have to be made in the future. A fairly extensive, though not comprehensive, bibliography is included to guide further reading.

We are grateful for the advice and comments we have received from many colleagues, and particularly to Dr A.J. Crane of the CEGB and Mr E.J. Allett of British Coal. The Oxford University Press has been most helpful and patient. We are glad to acknowledge the invaluable support of Mrs Christine Flack and Miss Juliet Smith in the preparation of the manuscript.

L.E.J.R.<br>P.S.L.<br>P.A.H.S.

August 1989

# Contents

# 1
## HISTORY AND GENERAL ENVIRONMENTAL IMPACT

Modern society takes few things so much for granted as the instant availability of almost limitless energy at the flick of a switch. All that is needed is a few strands of copper wire and, of course, a meter. Electricity enables us to enjoy a standard of living beyond the wildest dreams of our forefathers. We can have light and warmth wherever and whenever we need it, machinery can release us from domestic drudgery and the use of electricity is central to work and leisure. Some other forms of energy are more convenient for some purposes but none combines at the point of use high efficiency, immediate availability, absolute cleanliness, no storage requirements, silence, and absence of waste products.

But all benefits have to be balanced against costs and in the case of electricity the costs are the environmental effects of generation and distribution. Unlike most energy technologies, the environmental impacts of electricity are caused almost entirely by the producer rather than the user.

## The early years

There were few references to environmental considerations in the early years of the industry. 'Environmentalism' had not been invented and the generation of electricity added insignificantly to what was already a far from clean environment, dominated in towns and cities by smoke from the domestic use of coal. There were, nevertheless, some local problems. The St James's Electricity Company in London was prosecuted for discharging hot water into the drains in such quantities that the sewer men could not carry out their work and at Paddington there were complaints of 'tremendous vibration and noise, added to the fumes of smoke and steam (which) produced such a nuisance as to be almost unbearable' (Hinton 1979). A typical early installation would have been in the basement of a large building, with all the combustion products, including quantities of soot and dust, going straight up the chimney. The plants were not dissimilar to a steam locomotive.

The development of electric power dates from Faraday's discovery, in 1831, of electromagnetic induction—the discovery that a changing magnetic field could move a length of wire carrying an electric current. The original equipment can still be seen in the Royal Institution in London. The first continuous electric generator was demonstrated in Paris in 1832, but the driving force was a man turning a horseshoe magnet by hand. It took another 40 years to develop a practical machine. The first application was lighting. Rudimentary arc lights using banks of batteries and carbon electrodes had first been tried in 1802 but it was not until 1846 that carbon sufficiently pure and hard to produce a steady light became available. The most important early application was in lighthouses (Jarvis 1958). During the late 1870s a number of electric lighting installations were in use, in theatres, factories, and a few sites such as the Embankment in London. Early arc lighting was not without problems: at the first illuminated football match, at Sheffield, 'The brilliancy of the light . . . dazzled the players, and sometimes caused strange blunders' (Hennessey 1972). The carbon arcs had a short life; the gap had to be kept constant and the carbons frequently changed. A later development was the Jablockhoff candle which used a parallel pair of insulated carbon rods instead of bare rods end to end; the insulator burned away as the carbons burned. These also had their drawbacks and were difficult to keep going in a strong wind. Most of the early installations, which were more for show than for use, were short-lived because the cost was too high but some were sufficiently useful and survived well into the age of the incandescent lamp of Swan and Edison.

A large number of lighting systems were introduced during the early 1880s, in railway stations, hotels, and streets. The first public supply in the UK was at Godalming in Surrey in 1881. This was unusual in that it used hydropower, from the River Wey, and was thus virtually without environmental impact. Once again, the scheme was short-lived. There were not enough customers and it could not compete with gas—the town returned to gas lighting in 1884. The Brighton system, which started generating electricity early in 1882 was more typical of the many systems that were to spring up in the coming years, being driven by a coal-fired steam engine. The Brighton system grew rapidly, from supplying only 16 arc lamps from dusk till 11 p.m. daily in 1882 to powering 1000 incandescent lamps in 1886 and providing round-the-clock service in 1887 (Hennessey 1972).

After the early excitement and enthusiasm for a novel system had died down the main constraint on expansion was cost. Gas was in widespread use for lighting, heating, and cooking. Since electricity was used only for lighting, load factors were low and, except in some factories where lights were needed for long hours and there may have been spare engine power, electricity remained a luxury. In Britain it was not until 1911 that electricity became cheap enough to show a decisive advantage over gas for all new lighting installations (Byatt 1979).

A major expansion in electricity generation came with the development of a new market for electricity: trams and railways. The environmental problems associated with horse-drawn trams were minor compared with the total problem caused by urban horse-drawn traffic, and the replacement of the horse by an electric motor was not a major advance from an environmental point of view, although it may well have been from the point of view of the horse. The change from steam-driven to electrically-driven underground trains, however, was dramatic. Early underground lines consisting of short tunnels linked by open cuttings were unpleasant though bearable, but totally underground systems were clearly impractical with steam. Attempts to use moving wire ropes with fixed steam engines at each end were not successful. The first true 'tube', the City and South London railway passing under the Thames, was opened in 1890. This and the subsequent rapid expansion of the underground network in central London were based on the introduction of electrically-driven trains, and provide a good example of the shift of environmental effect from the point of use to the point of production.

The electrification of overground trains in cities and of inter-city trains was based on the comparative running and capital costs of electric and steam systems and not on the need for pollution-free engines. It is doubtful, however, whether large numbers of steam locomotives would now be acceptable in a central city terminus.

The third major market for electricity to emerge, after lighting and traction, was as motive power in manufacturing industry; indeed by 1912 factories used three times as much electricity in Britain as did traction and had overtaken the amount used for lighting (Byatt 1979). This, too, resulted both in safety and in environmental improvements, with the removal of complex systems of belts powered mainly by coal-fired steam engines. The major growth, however, came not because of these improvements but because of the adaptability, versatility, and rapidly falling costs of electricity.

As the market for electricity expanded and distribution systems grew from single buildings to streets to districts, the scale of generation needed became more and more difficult to accommodate. The obvious next stage was to move to a smaller number of central power stations to benefit from the economic advantages of scale and to supply a range of users, each with their different temporal demand patterns, from a single station, thus smoothing out the peaks in demand. One of the most ambitious projects of its time was Ferranti's plan to build a power station at Deptford 'in order to reduce the noise in Bond Street and gain access to cheap riverborne coal and cooling water' (Hannah 1979). Ferranti's Bond Street station consisted of 700-h.p. generators transmitting at 2500 volts. For Deptford he proposed 10 000-h.p. generators transmitting at 10 000 volts—a huge technological leap. Construction started in 1888 with 1250-h.p. units intended to operate until the larger units were built, but the scheme ran into technical and financial problems and was never completed. It was not until the 1890s that stations of a comparable scale were built and operated successfully, partly because of the development of the steam turbine to replace the earlier reciprocating engines, and partly because of the expansion of the market for electricity into transport and industry and the increased utilization of generating plant that became possible as a result.

Although the potential benefit of large-scale interconnection was great, it took many years for this to be achieved in Britain. Electricity was generated by users, by electric power companies, and by municipalities, all operating independently. The way forward was demonstrated by one of the largest of these, the Newcastle Electric Supply Company. This company provided electricity for a major industrial area, including big shipyards, a suburban railway system, collieries, and, of course, lighting; indeed it was the biggest integrated power system in Europe before the First World War (Hannah 1979). It took the pressures of that war on manufacturing industry to produce further improvements—interconnections and standardization became more common and average load factors increased.

After the war, realization of the benefits of these improvements and of the importance of electricity supply led, in Britain, to the appointment of Electricity Commissioners who tried, without much success, to introduce rationalization on a voluntary basis. One of the problems, particularly in London, was the enormous variety of systems that had been developed. In 1913, for example, Greater London had sixty-five electrical utilities, seventy generating stations, forty-nine different types of supply systems, ten different frequencies, thirty-two voltage levels for transmission and twenty-four for distribution, while Chicago and Berlin each had a centralized system (Hughes 1983). The critical step was the setting-up of the Weir Committee in 1925 which made strong recommendations which were endorsed by the government. The Electricity (Supply) Act 1926 set up a Central Electricity Board charged with the job of building a national grid and exercising overall control of electricity generation. Planning and construction of the grid progressed rapidly but was not without problems. The impacts of industrialization had previously been essentially in towns and cities. The grid was a major intrusion into otherwise untouched countryside and the problems were similar to those caused by the introduction of railways. The situation was made worse by the fact that the benefits of electricity were generally not available to those most affected— the landowners who had to put up with the visual intrusion and inconvenience of pylons on their land but who were not offered access to a supply of electricity. A not untypical reaction was 'You mean I'll have to look at that monstrosity passing right in front of my windows, spoiling my best views, interfering with my grazing and ploughing and harvesting, and I won't even get any electricity out of it?' (Cochrane 1985). In some particularly sensitive areas such as the Lake District, the Sussex Downs, and the New Forest, cables were routed to avoid the finest beauty spots, or placed underground.

This early example of the difficulty of balancing national benefit against local loss of amenity was to become an important factor in later developments, and has become particularly acute recently in the debate about the disposal of radioactive wastes from the generation of electricity by nuclear power.

The grid was essentially complete and in full commercial operation by 1935, and the economic benefits were immediate. It became possible to increase overall efficiency by making most use of the most efficient stations according to a 'merit order', and the amount of reserve plant needed to satisfy peak load requirements was reduced from 75 per cent to 15 per cent (Cochrane 1985).

Although the grid offered the opportunity of removing the main source of pollution, i.e. smoke, from urban to isolated rural areas, this was not taken—had it been, opposition from landowners would probably have been great. The grid was seen essentially as a way of interconnecting the main industrial areas and providing some links between them to enable them to help each other out in emergencies or when maintenance was needed. The capacity of the inter-regional links was small. The only direct environmental advantage was the ability to make national use of hydropower sources for which there was no local load. As a result the contribution of hydropower in Britain increased from virtually nothing in 1918 to 4.2% of electricity generated in 1938 (Hannah 1979). Power stations were still being built within or very close to cities rather than near to coalfields. As they increased in size, there were growing difficulties with environmental impact, both visually and because of pollution from smoke. One of the most sensitive installations, being within sight of the Houses of Parliament, was the Battersea power station, approved in October 1927. The consent contained a condition that the Company should take the best known precautions 'for preventing as far as reasonably practicable the evolution of oxides of sulphur, and generally for preventing any nuisance arising from the generating station'. The publicity material for the station describes the measures taken:

Coals having a low sulphur content are used exclusively at Battersea. The flue gases from the induced draught fans pass to the 'primary' chambers where the gases are first water-sprayed in contact with steel channels. They then pass into the main flue connecting the two chimneys where contact is made with similar channels. Arriving at the chimney towers the gases run through a central downtake which is also equipped with sprays and steel channels. The gases are thus subjected to more or less continuous spraying which, assisted by the catalytic action of the iron oxide present, converts the $SO_2$ into $H_2SO_4$. The gases subsequently ascend the two side uptakes of the towers where they make contact with wooden scrubbers and receive an alkaline wash, finally passing through wooden moisture eliminators to the chimneys after receiving an admixture of hot air at the base of the latter. This gas washing is a unique feature of the station. Not less than 90 per cent of the sulphur oxides contained in the flue gases are eliminated by this process. (London Power Company, undated).

It took until 1987 for sulphur dioxide removal to become a requirement in all new coal-fired power stations in Britain. However, there has long been a requirement to provide for the possibility of fitting removal equipment. For example, the Ministry of Power consent for the Drax power station, approved in September 1964, included the condition 'The layout of the Station shall be designed so as to permit the installation of such plant as may be reasonably practicable for the prevention of the discharge of sulphur and its compounds into the atmosphere, and the Board shall, if so required at any time by the Minister, install such plant.'.

The first demonstration that the grid could be operated as a unified system rather than as a series of links between local networks was made in October 1937, on the initiative of a control engineer in the south-eastern control room in London. 'Quite unofficially, the control engineer on shift issued the switching instructions; one by one the seven areas were all coupled—and it worked. Every power station in the country which was connected to the grid that night was operating for the first time as part of one completely integrated system,' (Cochrane 1985). The full benefit of such a system was soon to be needed. Meeting the rapidly expanding needs of war industries and maintaining supplies in spite of bomb damage was a major challenge. The grid system, controlled from bomb-proof underground bunkers in London, enabled vital supplies to be maintained throughout the war.

## Post-war developments

Ironically, the problems of the immediate post-war years were more difficult to cope with than the problems of the war years. Lack of adequate maintenance during the war, inadequate construction programmes and coal shortages resulted in load-shedding and power cuts. The most serious crisis was during the exceptionally severe winter of 1947, with only ten days supply of coal at the power stations and some of that frozen too hard to use. Electricity use was only allowed for essential industries and emergencies, factories closed down, domestic and commercial use was restricted, electric trains were cut, and the country shivered.

Nineteen-forty-seven was also the year in which electricity in Britain was nationalized. A major construction programme was launched to meet the growth in demand that came with post-war recovery and with the growth in electrification. In 1947 one-quarter of British households still had no electricity: this was reduced to less than one-tenth by 1957. It took some years, however, for supply to catch up with demand, and power cuts remained a feature of life during the winter until the early 1950s.

From 1950 onwards the size of power stations gradually increased. In the early years of nationalization, 30-MW and 60-MW sets, the same size as before the war, were ordered. In 1950 the first 100-MW set was ordered; 200-MW and 275-MW sets were ordered in 1959 and 1962, and 500-MW sets came into use later in the 1960s—not without considerable teething troubles. In the 1970s 600-MW sets were introduced with fewer problems.

With the increase in unit size came lower costs and the development of a 'supergrid' at 275 kV, with a much higher power carrying capacity than the original 132-kV grid. Many of the power station sites inherited from pre-nationalization days were too close to centres of population for the larger power stations and the reduction in transmission costs that came with the supergrid paved the way to the final removal of direct environmental impact from the areas of high electricity use. The transport of electricity became cheaper than the transport of fuel—'coal by wire'—and large power stations were built mainly near coal-fields and oil refineries or, in the case of the nuclear power stations introduced from the 1960s onwards, on the coast (Fig. 1.1). Parts of the grid were upgraded to 400 kV and by the early 1970s total generating capacity had grown to nearly 50 Gigawatts—four times the capacity at the end of World War II.

Proposals for the privatization of the electricity supply industry in the UK are currently before Parliament.

## General environmental considerations

The unprecedented rate of innovation in science and technology that has occurred this century, and in particular since World War II, has included striking developments in the range, sensitivity, and precision of methods of measuring environmental pollutants. The very availability of information on pollutants has resulted in deeper questioning of the acceptability of the technologies that produce them, sometimes to the neglect of considerations of the overall benefits resulting from the availability of the technologies themselves. The rate of introduction of new products or techniques in the food and chemical industries, in agriculture and horticulture and in the energy industries is now strongly influenced by their actual or perceived environmental effects.

While the remainder of this book is about various ways in which the generation of electricity may have harmful effects on the environment, it is important not to lose sight of the enormous benefits that electricity brings. Electricity has played a key role in the scientific, technological, and medical developments of the past century that have resulted, at least in the

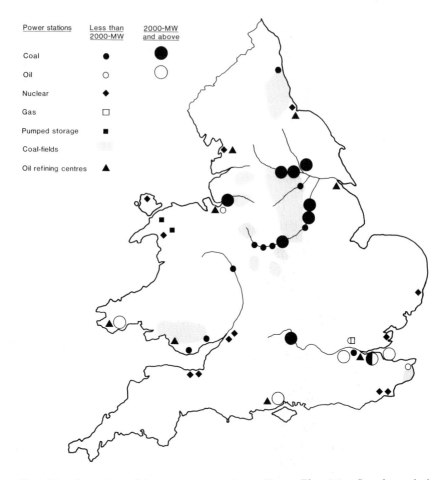

**Fig. 1.1.** Location of large power stations. From *Electricity Supply and the Environment* (CEGB).

more developed countries, in unprecedented improvements in health and in other measures of physical wellbeing. There is a close correlation between such measures as infant survival rates and overall expectation of life and energy consumption (which is closely related to electricity consumption) per head (Fig. 1.2), and similar patterns are found for other measures of living standards. All the major sources of energy have, by raising our material standards of living, lengthened far more lives than they have shortened (Fremlin 1985).

The preservation of amenity has been a formal requirement of the electricity generating industry since the Electricity Act (1957), Section 37 of which states:

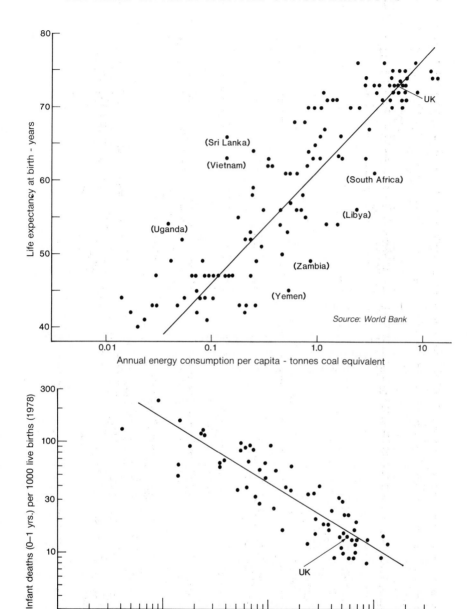

**Fig. 1.2a**   Correlation of life expectancy with energy consumption per head. (121 countries)

**Fig. 1.2b**   Correlation of infant mortality with energy consumption per head. (65 countries)

In formulating or considering any proposals relating to the functions of the Generating Board or of any of the Area Boards, . . . the Board in question, the Electricity Council and the Minister, having regard to the desirability of preserving natural beauty, of conserving flora, fauna and geological or physiographical features of special interest, and of protecting buildings and other objects of architectural or historic interest, shall each take into account any effect which the proposals would have on the natural beauty of the countryside or on any such flora, fauna, features, buildings or objects.

There is no specific mention of human health, now generally accepted as a key consideration in assessing environmental impact. Part 1 Section 1 of the Electricity Act (1947) does, however, require the Electricity Boards to 'promote the welfare, health, and safety of persons in the employment of the Boards'.

The specific environmental impacts of fossil, nuclear, and alternative forms of power generation will be dealt with in subsequent chapters. There are, however, a number of factors that are common to all methods of generating and using electricity on a large scale. These are land and water use, visual intrusion, thermal pollution (except for hydropower), and transmission and distribution.

The land requirements of a power station are small: typically 60 hectares for a nuclear station, including the construction area, and about twice as much for a coal-fired station because of the need to store coal. If ash is disposed of locally considerably larger areas are needed. The land has to be able to support very heavy loads and should not be too far above the source of cooling water to avoid excessive pumping costs. While the best agricultural land or sites of particular natural or scientific interest are avoided wherever possible, a power station need not limit farming on adjacent land.

Very large quantities of cooling water are needed both for fossil and for nuclear stations. A direct-cooled station of 2000 MW (electrical) would need about 60 cubic metres of water per second. In general this is available only at coastal sites. Inland sites need cooling towers; at a typical station these would need about 2 cubic metres of water per second to replace evaporation losses and as purging water to prevent excessive build-up of dissolved substances. The extraction of the water may result in some damage to fish and other creatures but only in a small area immediately around the intake.

Visual intrusion is an inevitable consequence of the scale of modern power stations. A typical inland station needs six to eight cooling towers each 110 metres high and 90 metres in diameter at the base. The building housing the boilers (coal or nuclear) is about 200 metres long and 60 metres high. A fossil-fired station has a smoke stack up to 200 metres high. To some, such massive structures have a beauty of their own. The 2000-MW coal-fired station at West Burton in Nottinghamshire (Fig. 1.3) received an

**Fig. 1.3**    The West Burton coal-fired station (CEGB).

award from the Civic Trust, which described it as 'an immense engineering work of great style, which far from detracting from the visual scene acts as a magnet to the eye from many parts of the Trent Valley and from several miles around' (CEGB 1979). To others their very visibility makes them a massive eyesore. The creation of artificial hills and tree planting can help but the stations inevitably become dominant features in the landscape. Nuclear stations present less of a problem because they need no large coal stores or oil tanks and have no tall chimneys. Since they do not need to be near to railway lines, ports, or refineries more sites are in principle suitable. Coastal siting avoids the need for cooling towers, further reducing visual impact.

A power station with direct cooling ejects about two-thirds of the total heat generated in the form of a massive flow of water heated to 10–12°C above the intake temperature. There has been much concern that this 'thermal pollution' would cause serious damage to the local ecology. In practice the warm turbulent water seems to attract fish, which in their turn attract fishermen, seabirds, and ornithologists. There are a number of reasons for the lack of harmful effects: the flow of cooling water only slightly perturbs the huge natural flows at typical coastal or estuarial sites; the warm water floats in a thin layer on the surface of the main body of water so that it has no effect on sea-bed life and dissipates its heat rapidly to the atmosphere; the natural ecology is already able to tolerate wide swings in temperature caused by seasonal and daily changes; and the turbulence in the flowing water results in aeration that compensates for oxygen losses due to the temperature increase (CEGB 1979).

The transmission and distribution of electricity to virtually every home, office, and factory in the country requires a network not unlike the blood circulation system of the human body with the additional complications of several independent 'hearts' and of switching and voltage conversion at many of the junctions. The most obvious impact is the visual one of the tall towers of the super-grid, but there are also problems with distribution at lower voltages, with the siting of substations housing switchgear and transformers, and with the noise that these components can produce. The very high cost of placing high-voltage transmission lines underground— 10 to 20 times that of overhead lines—makes the total disappearance of the grid unlikely unless there is a technical breakthrough, for example in the field of superconductivity. Much progress has been made, however, at the lower voltage end of the system with the great majority of transmission below 11 kV being underground. Noise problems can be largely overcome by the use of acoustic enclosures and silencers for high-voltage switchgear.

The final destination of the electricity is associated with the least environmental impact and the greatest benefit over alternative sources of energy. The unique advantages of electricity to the user, stemming from its efficiency, versatility, and cleanliness and from the steady fall, in real terms, in its price have led to its becoming, at least in developed countries, the fastest-growing form of energy and the one to which the largest number of people have access. We cannot, however, ignore the risk of electrocution, in the home and in industry, which in 1985 accounted for 65 out of the total of 13,715 accidental deaths in England and Wales (OPCS 1987). Over half of the deaths from electrocution occurred in the home. We can, however, definitely discount the fears of James Thurber's maternal grandmother who 'lived the latter years of her life in the horrible suspicion that electricity was dripping invisibly all over the house. It leaked, she contended, out of empty sockets if the wall switch had been left on.' (Thurber 1945).

## Planning considerations

We have seen the environmental benefits that have resulted from the separation of electricity generation and use and these suggest that power stations should be sited in the most remote locations available. In practice, however, there are a number of constraints. Long transmission lines are expensive, result in loss of power, and are themselves visually intrusive; moreover, remote locations are often valued for their very remoteness in a densely populated country such as Britain. It is generally both more efficient and less environmentally intrusive overall to have a reasonable regional balance between generation and demand.

In Britain, electricity demand is at present growing most rapidly in the south and the transmission links to the Midlands and North are approaching their limit (Gammon 1987). Sites for new power stations are therefore being sought in the southern half of the country, which is already the most developed and which already has large areas protected for environmental reasons. The planning problems, both for coal and for nuclear stations, are likely to be considerable.

The obvious approach is to extend existing power station sites: grid connections and water supplies are already available and the local population is at least familiar with any problems that may exist, even if it is not always unconcerned about them. The Central Electricity Generating Board (CEGB) is using this approach, for example at Sizewell and Hinkley Point for nuclear stations and at West Burton for coal-fired stations, and is likely to continue to do so wherever possible. The identification of new sites is considerably more difficult.

The problem is well illustrated by the situation in the south-west Peninsular, the counties of Somerset, Dorset, Devon, and Cornwall. Much of the area is outstandingly beautiful and there is a large and economically important tourist trade. More generation capacity is urgently needed in the area and intensive searches for suitable sites have been made for many years (Gammon 1987). Few were found that even merited detailed study. The three deemed to be the most promising were a further nuclear station at Hinkley Point in Somerset, a nuclear station near the United Kingdom Atomic Energy Authority's (UKAEA) research establishment at Winfrith in Dorset, and a coal-fired station near Plymouth. The CEGB is currently pursuing the Hinkley Point option, on which a major Public Inquiry opened in 1988. In 1989 the CEGB announced that it was intending to seek approval for two further PWR stations, one at Wylfa and a further one at Sizewell, and that it was not proceeding with plans for a coal-fired station at Fawley on Southampton water.

The CEGB has summarized its planning approach to satisfying the environmental constraints on development of power stations as follows (Gammon 1987):

When implementing system strategy:
(1) to locate power stations where they reduce the need for new long-transmission lines by keeping a broad balance between electricity demand and generation in geographic areas;
(2) to give regard to conserving inland water resources.

When selecting sites
(1) to extend or redevelop existing sites to fully use existing facilities and to reduce the number of greenfield sites;
(2) to accept some areas of national or international importance as being sacrosanct;

(3) to avoid as far as possible new sites in areas given protection because of their environmental value;
(4) where sites are developed in protected areas to give special attention to reduce environmental effects and to give a degree of compensation through creative conservation;
(5) to make judgements of the comparative balance of environmental and economic factors for the alternatives available.

On project development

(1) to take account of the environmental issues when optimising layout and plant performances:
(2) to monitor the environmental effects of existing plant and feed back information to improve the design of new stations;
(3) to find ways of reducing significant environmental effects, of ameliorating them and of giving compensation.

During the past few years there has been an increasing, and welcome, public and political concern with environmental issues, particularly with those relating to power generation such as the problems of acid rain and the greenhouse effect. The next few years will show the extent to which electricity industries world-wide will be able to satisfy the joint requirements of providing an economic and reliable electricity supply and meeting their environmental obligations.

## References

Byatt, I.C.R. (1979). *The British electrical industry 1875–1914*. Clarendon Press.
CEGB (1979). *Electricity supply and the environment*.
Cochrane, R. (1985). *Power to the people*. Newnes.
Fremlin, J.H. (1985). *Power production: What are the risks?* Adam Hilger.
Gammon, K.M. (1987). *Planning the development of electricity generating systems*. CEGB Newsletter No. 1.
Hannah, L. (1979). *Electricity before nationalization*. Macmillan.
Hennessey, R.A.S. (1972). *The electric revolution*. Oriel Press.
Hinton, Lord (1979). *Heavy current electricity in the United Kingdom*. Pergamon Press.
Hughes, T.P. (1983). *Networks of power*. Johns Hopkins University Press.
Jarvis, C.M. (1958). Chapter 10 *A history of technology* (ed. C. Singer *et al.*) Vol. 5, Ch. 10. Clarendon Press.
London Power Co. Ltd. brochure (undated). *The Battersea power station*.
OPCS (1987). *Mortality Statistics*.
Thurber, J. (1945). *The Thurber carnival*. Penguin.

# 2

# GENERATION FROM FOSSIL FUELS

The fossil fuels, coal, oil, and gas, now provide about 60% of the world's electricity and are likely to remain important fuels for electricity generation for many decades (hydro power provides about 23% and nuclear power about 17%). The relative contributions of these fuels have changed dramatically since the early days, when almost all electricity was generated from coal. In 1965, for example, 44% of world electricity (excluding the communist countries) was generated from coal, 12.5% from oil, and 10% from gas. The corresponding figures for 1985 were 37%, 11%, and 10.5%. The total quantity of fossil fuels burned each year for the generation of electricity (again excluding the communist countries) increased from 1105 million tonnes of coal (or coal equivalent) in 1965 to 2975 million tonnes in 1985.

Although a large amount of electricity is still generated from oil and gas, the other markets for these fuels are much larger and only a fraction of the environmental impacts of their extraction and distribution should therefore be apportioned to electricity generation. Coal, on the other hand, is now used more for the generation of electricity than for any other purpose— about 80% of domestic coal production in Britain, for example, compared with 1.7% in 1913, 7% in 1939 and 15% in 1947. In assessing the overall environmental impact of coal-fired electricity generation, therefore, it is reasonable to include the impacts of extraction and distribution as well as of coal burning and ash disposal.

The environmental impact of electricity generation from all fossil fuels is due largely to their relatively low energy density compared with nuclear power although, as we shall see in Chapter 4, the energy density of renewables is still lower and this is an even more important factor governing their environmental impact. In order to generate one giga-watt year of electricity (1 GW(e) yr) or the electricity needed to run one million one-kilowatt electric heaters for one year, one needs to mine and transport 3.8 million tonnes of coal, or extract and transport 2.2 million tonnes of oil. In 1985 74 million tonnes of coal were burned in Britain to generate electricity; the amounts of gaseous and solid wastes to be dealt with are therefore very large.

## Oil

The world currently extracts, transports, and consumes nearly 3000 million tonnes of oil a year. The main direct environmental impacts of this vast enterprise come from the fraction that inevitably gets lost by spillage or accident. Blowouts and spills at wells result in losses of only 0.01 to 0.02 per cent of output, but this adds up to 300 to 600 thousand tonnes of oil a year (OECD 1985). Most oil is transported by tankers and about 2 million tonnes a year enter the marine environment as a result of routine operations and accidents. Apart from the obvious unpleasantness of oil on beaches and the economic impacts on tourism and commercial fishing, the main effects are on wildlife. The loss of the Amoco Cadiz in 1978 for example resulted in the spillage of over 200 000 tonnes of oil; serious damage to the local ecology occurred over much of the north-west coast of Brittany, including the deaths of several hundred thousand tonnes of sea creatures and 15 000 to 20 000 birds. (Lagadec 1982). The full ecological consequences of the major spillage from the Exxon Valdez in Prince William Sound, Alaska, in 1989 are still being assessed.

One of the most serious tanker accidents of recent years was the explosion of the Betelgeuse in Bantry Bay, Ireland, in 1979; this caused 48 deaths. The accident could have been very much more serious because of the presence of a nearby tank farm containing several million gallons of petrol. (Lagadec 1982).

The operation of oil refineries results in the emission of several thousand tonnes a year of airborne and liquid effluents. Table 2.1 lists the effluents from a refinery that processes about 20 million tonnes of crude oil a year (OECD 1985). The environmental effects of these are essentially the same as those of similar emissions from coal-fired power stations.

In addition to the effects of routine refinery operation, there is the risk of large-scale accidents. The Health and Safety Executive (HSE) in Britain has assessed the risks associated with the Canvey Island complex on the Thames estuary, which contains two refineries. The assessment is complicated by the presence of other hazardous plant such as a liquid natural gas (LNG) store and an ammonia store, but it is possible to estimate a risk that can be apportioned to the percentage of the total operation that would be associated with the generation of 1 GW(e) yr in an oil-fired power station. This is shown in Table 2.2 (Cohen and Pritchard 1980); it is significantly higher than the risk of similar numbers of deaths resulting from major nuclear accidents, discussed in Chapter 3.

The environmental impacts of burning oil to generate electricity do not differ greatly from those of burning coal (see p. 22) except that particulate emissions are considerably smaller and there is no ash to dispose of.

**Table 2.1**   Effluents from an oil refinery*

|  | Metric tonnes per year |
|---|---|
| *Airborne effluents* |  |
| $SO_x$ | 21 000 |
| Organic compounds | 22 700 |
| $NO_x$ | 17 700 |
| CO | 4300 |
| Ammonia | 2200 |
| Particulates | 2 800 |
| *Liquid effluents* $(1.4 \times 10^8$ tonne waste water) |  |
| Containing: |  |
| Chlorides | 24 000 |
| Oil | 600 |
| Ammonia nitrogen | 600 |
| Phosphate | 600 |
| Suspended solids | 3 |
| Dissolved solids | 2000 |
| Trace metals (Cr, Pb, Zn, Cu) | 22 |

*Refinery processes $2.3 \times 10^7$ tonnes crude oil per year. *Source: OECD (1985)*

**Table 2.2**   Estimate of associated refinery risk linked to output of one 1000 MW(e) oil-burning power station, 75% load factor. Annual chance per million, assuming all recommended safety precautions taken, of an incident causing more than 1500, 4500, and 18 000 deaths.

|  | 1500 deaths | 4500 deaths | 18 000 deaths |
|---|---|---|---|
| Including ammonia store | 18 | 7 | 1.6 |
| Excluding ammonia store | 11 | 4 | 0.4 |

*Source: Cohen and Pritchard (1980)*

## Gas

Gas is often produced in conjunction with oil and the apportionment of emissions and accident data between the two depend on the fuel composition of the reservoir and the nature of the operation (offshore or onshore) (OECD 1985). The routine environmental impacts associated with gas production and distribution in gaseous form are small. The risk of accidents associated with the transport and storage of liquefied gas, however, is significant and most of the serious energy-related accidents that have occurred during the past few decades have involved liquefied gases. Table 2.3 lists some of the most serious.

**Table 2.3**  Accidents involving liquid gases

| | | | |
|---|---|---|---|
| Cleveland, USA | LNG | 1944 | 136 deaths |
| Feyzin, France | Propane | 1966 | 16 deaths |
| Los Alfaques, Spain | Propylene | 1978 | 216 deaths |
| Xilatopec, Mexico | LNG | 1978 | 100 deaths |
| Cubatao, Brazil | LPG | 1984 | over 500 deaths |
| Mexico City, Mexico | LPG | 1984 | over 500 deaths |

## Coal

### Coal extraction

The popular image of coal still largely reflects the past: soot-encrusted buildings, dereliction, and spoil heaps in old coal-mining areas, subsidence, and high accident and disease rates among coal-miners. While coal is unlikely to be perceived as a clean fuel compared, for example, with electricity, oil, or gas, today's coal industry, in most countries, is dramatically different from that of the past. Nevertheless, the enormous quantities of material that have to be extracted and moved have inevitable environmental effects.

The largest direct effect is probably that of opencast mining. The scale of operation can be vast: in Germany, for example, there are lignite mines where areas exceeding 20 km$^2$ are dug out to depths of up to 400 metres (OECD 1985). Earth and rock moving on such a scale cannot be done without major problems of dust, noise, and vibration. More serious in the long term may be permanent damage to the original top-soil structure, local ecology and surface and groundwaters. In the past, huge areas have become virtually sterile, for example in the Appalachians in America as a result of strip-mining in hilly country.

In Britain, opencast mining is on a much smaller scale than that, for example, in Germany and America; strict standards are applied to all stages of the operation and subsequent restoration to minimize environmental impact. Opencast production has stayed about constant over the past thirty years at 12–15 million tonnes a year. British Coal's opencast operations now cover about 12 500 hectares of land with about 7500 hectares of previously worked land being restored. In order to maintain production at the current level some 2000 hectares of land need to be brought into production annually (HMSO 1987). The environmental impacts of these operations were reviewed by the Commission on Energy and the Environment in 1981 (HMSO 1981) and by the House of Commons Energy Committee in 1986–1987 (HMSO 1987). The latter agreed with the view that 'opencast mining is one of the most environmentally destructive processes being carried out in the UK' but concluded: 'Provided the full range of environmental costs are properly acknowledged in any evaluation of a potential opencast site, and provided the economic benefits are examined with similar rigour, if the benefits exceed the costs, we see both a commercial advantage to British Coal and other operators, and an economic advantage to the nation, from maintaining opencast operations at around the present level.' The Government's response to the 1981 report of the Commission on Energy and the Environment stated that 'the new arrangements for dealing with opencast planning applications will ensure that the Board (now British Coal) continue to carry out and develop further their high standards of restoration and after-care work.' (HMSO 1983).

Deep coal mining does not in general have the obvious and direct effects on the environment that are inevitable with opencast mining, although in the past the industry has been responsible for a great deal of dereliction—mainly in the form of spoil heaps. Drainage water from a deep mine has caused serious problems in some countries; in America, for example, about 16 000 km of streams and rivers, principally in Appalachia, have been degraded by acid waters from mines and in Germany, drainage from coal mines has contributed to the salinity of the Rhine (OECD 1983). In Britain this has not been a serious problem in spite of the very large quantities of water that have to be pumped from mines—an average of 2.3 tonnes of water per tonne of saleable coal with as much as 30 tonnes of water per tonne of saleable coal in some mines (HMSO 1981).

The two most important environmental problems associated with deep mining are spoil disposal and the possibility of subsidence. For every tonne of saleable coal produced about half a tonne of spoil has to be separated and disposed of; the basic problem is the huge volumes of material that have to be dealt with. The Commission on Energy and the Environment concluded that spoil would be one of the major environmental problems arising from deep mining in the 20 years following their report (HMSO 1981). Every 50 million tonnes of spoil requires about 200 hectares of new land. There have been very striking improvements since the early days of the industry, particularly following the Aberfan disaster of 1966 when 144 people, mostly children, were killed when an old spoil heap became unstable and slid into a town, engulfing houses and the local school. The causes of this disaster are now well understood and modern waste tips are carefully sited, well compacted, designed to have a low profile, and progressively restored and landscaped. Britain now contains examples both of derelict old spoil heaps still to be restored and successful restoration schemes with contoured new hills which can enhance the local environment once the operation is completed.

Some degree of subsidence is unavoidable for both main methods of deep mining coal—long-wall and room and pillar—but subsidence does not necessarily lead to damage.

In long-wall mining, as in Britain, subsidence occurs almost immediately as the hydraulic supports that hold up the roof while cutting takes place are moved forward to a new position. The amount of subsidence can be predicted quite accurately. The area affected is larger than the area from which coal is extracted, and depends on the depth of the workings— typically the area of the workings plus a distance from vertically above the edge of the workings of 0.7 times the depth is affected. The amount of subsidence also depends on the depth of the workings and, of course, on the thickness of the seam being extracted. Subsidence is typically from a few centimetres to 0.5 metres and the damage ranges from very slight to severe. In one particular area in Britain that was studied in detail, about a third of the houses in the area suffered some damage. Of the 62 500 houses damaged, however, few (about 2000) suffered 'appreciable damage' and even fewer (about 250) suffered 'severe damage' (HMSO 1981). The area in question was known to have a higher susceptibility to subsidence-related damage claims than anywhere else in Britain. Compensation and repair payments for subsidence damage in Britain are currently running at £90 million a year (HMSO 1987). Subsidence can also affect agricultural land, roads, gas, sewage and water pipes, and local hydrology.

Room and pillar mining results in less subsidence but the subsidence that does occur is difficult to predict and may continue over a long period. About 25 per cent of the 30 000 km$^2$ of undermined land in the Eastern United States has suffered some subsidence (OECD 1983).

### Coal transport

Coal is transported by rail, road, sea, canal, or conveyor belt. In Britain, for example, about three quarters of CEGB's coal is transported by rail, 13% by road, 8% by sea, 3% by canal, and 1% by conveyor belt (CEGB 1987).

Rail transport is mostly by merry-go-round trains. The trains are loaded and unloaded while on the move, minimizing the need for shunting and marshalling yards. Cohen and Pritchard (1980) have considered the apportionment of total rail transport accidents to coal-fired electricity generation and estimated that in Britain the operation of a 1000 MW(e) station at 75 per cent load factor would be associated with 0.03 occupational and 0.2 public deaths per year. Higher figures are estimated for countries like the USA and Canada where transport distances are generally longer and there are more level crossings.

## Ash Disposal

A typical large coal-fired power station (2000 MW(e)) produces over 2000 tonnes of ash a day, most of which is very fine pulverised fuel ash (PFA) (HMSO 1981). Some of this is used commercially, mainly for roadfill and building materials; the rest has to be disposed of. The CEGB currently produces about 12 million tonnes of ash a year, of which just under half is sold or used.

PFA is an organic mineral rich in aluminia, silica, and calcium, magnesium and iron oxides. It also contains a very wide range of trace elements such as lead, arsenic, selenium, mercury, copper, nickel, and zinc. Concentrations of most of these materials vary considerably depending on the source of the coal: arsenic 2–500 $\mu$g/g, mercury 0.02–0.4$\mu$g/g, lead 5–1000 $\mu$g/g and zinc 50–5000 $\mu$g/g (Natusch 1978).

PFA also contains radioactive materials. The naturally occurring radionuclides present at low levels in coal tend to concentrate in the solid residues and the specific activity of coal ashes is typically 2–15 times greater than that of the original coal. The resulting levels are still generally within the range found in natural soils. PFA also contains traces of a range of carcinogenic materials such as polycyclic hydrocarbons. These are further discussed later in this chapter.

Waste PFA is moved by road, rail, conveyor, or pipeline to landfill sites, close to power station sites whenever possible. These sites generally take only PFA but in a few locations PFA is used as a cover material at household waste landfill sites. Potential dust problems during transport and disposal have to be dealt with by the use of water sprays, prompt covering of exposed surfaces with soil, sprays of polyvinyl binding material, and rapid revegetation.

The main environmental problems associated with ash disposal are loss of land and amenity and potential water pollution, and air pollution resulting from dust blow. Of these the most serious potential hazard is probably the leaching of heavy metals into surface and groundwaters. Elements such as arsenic, cadmium, and selenium tend to occur at the surfaces of the particles and are more readily released than elements that occur predominantly within the glassy ash matrix. The great majority of the constituents of PFA are insoluble and only a fraction of the soluble materials are potentially harmful. The high alkalinity of the water in contact with the ash inhibits the solution of many heavy metals. Experiments by the CEGB have shown that only a few potentially toxic elements occur in the initial leachate from disposal sites in concentrations that exceed the World Health Organization recommended limits for drinking water (Clarke 1981). These may include chromium and selenium but the concentrations fall by a factor of ten as percolation proceeds. The leachate is subsequently diluted as it moves through the surrounding soil and adsorption processes are also likely to retard the movement of the potentially hazardous materials. The situation is akin to that in radioactive waste disposal sites (Chapter 3). The level of understanding, however, about the quantities of material involved, the detailed pathways through the surrounding rocks and the biosphere and the mechanisms of uptake and concentration in crops, animals, and humans is rudimentary compared to that about radioactive substances, and in general no estimates are available of the possible long-term hazards from toxic and carcinogenic materials in PFA.

## Atmospheric emissions from burning fossil fuels

In considering emissions to the atmosphere from the burning of fossil fuels it is first necessary to take a broad look at the chemical composition of the fuel itself, whether oil, gas, or coal.

### Composition of oil

Crude oil as pumped from the ground is a complex mixture of chemicals with an average elemental composition of 85–90% carbon, 10–14% hydrogen, 0.2–3.0% (or more) sulphur, and traces of many other elements including nitrogen, vanadium, and nickel. Structurally these elements are in the form of straight and branched chain paraffins (compounds containing only carbon and hydrogen). There are also carbon ring structures, such as cyclo-paraffins and aromatics, and it is in these types of compounds that the sulphur and nitrogen atoms generally occur. Before use the crude oil is separated into various fractions (by distillation, termed 'cracking'), and it is the heaviest fraction (boiling point greater than 350°C) which is used in power stations.

### Composition of gas

The lightest fractions of crude oil are gases at room temperature and consist of the simple paraffin molecules methane ($CH_4$), propane ($C_3H_8$), and butane ($C_4H_{10}$); the latter two can be liquefied at only moderate pressures and are known collectively as liquefied petroleum gas (LPG). These natural gases burn very cleanly and are used in industrial processes demanding a non-smoky flame, and domestically. Their high cost means that they are little used for power generation. Because of this, and in view of the cleanliness with which they combust, there will be no reason to refer to them specifically in later sections where power station emissions are discussed.

## Composition of coal

Coal is a mixture of organic compounds of high molecular weight and complex structure. Unlike oil, many of the organic compounds in coal have not been structurally identified. The elemental composition is within the following approximate ranges: 70–90% carbon, 4–5% hydrogen, 1–2% nitrogen, 5–15% oxygen, 0.5–5.0% sulphur, as well as many other elements in trace amounts, including chlorine, heavy metals (such as iron, cadmium, mercury, and zinc) and some radioactive substances. This composition varies from region to region, as well as between mines in the same area. Quite a lot of the variation comes about from differences in the time and conditions of temperature and pressure under which the coal has been formed during burial. As these factors increase, the ratio of carbon to the more volatile constituents, oxygen, hydrogen, and nitrogen also increases. As the proportion of carbon gets larger the 'rank' is said to get higher, so that the coal passes along a series from peat (low rank) to lignite (brown coal), through bituminous coal to high ranking anthracite. The origin of the sulphur in coal is not completely understood but in low-rank samples it occurs mostly in organic molecules, whereas in older, higher-ranking bituminous coals about half of the sulphur occurs as pyrite ($FeS_2$) or other inorganic compounds.

## Combustion of fossil fuels in power stations

The prime objective in a conventional power station is to burn the fossil fuel, whether it be oil, coal or gas, as efficiently as possible, so that the maximum amount of energy is obtained. With modern furnace design the efficiency with which the chemical energy in the fuel is converted to heat energy is very high (generally greater than 90 per cent). In passing, it is worth recalling that the reason why the overall conversion efficiency of conventional power plants is only 30–40 per cent is because of the fundamental inefficiency of the conversion of heat to electricity via the steam turbine.

Figure 2.1 shows the essential components of the combustion system in a fossil fuel power station. The diagram is for a coal-fired system—the only significant differences where the fuel is oil are the absence of the coal bunker and pulveriser and the ash hoppers. Also, in an oil-fired plant the electrostatic precipitator is replaced by cyclonic grit arrestors, the function of either device being to remove fine particles from flue gases. This

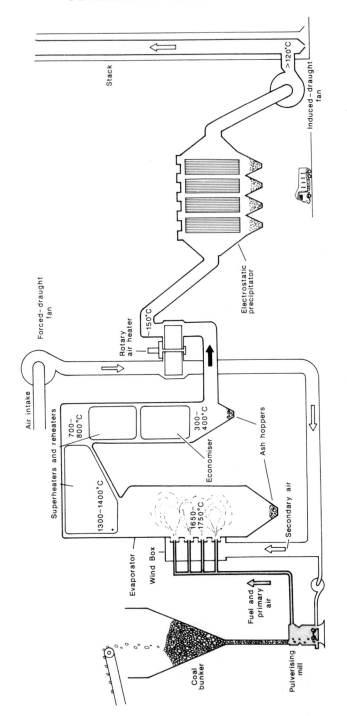

**Fig. 2.1** Schematic combustion system of a coal-fired power station (from Hart and Lawn 1977).

function is a very important one from the point of view of environmental pollution, since the particles removed from the gas stream contain non-volatile inorganic matter which contains a large proportion of most of the metallic elements in the original fuel; one or two exceptions to this general rule are discussed later. Electrostatic precipitators work by passing the stream of gas plus particles through a system of discharge wires and collector plates charged to a high potential. The wires ionize and hence charge some of the gas molecules which attach themselves to the particles. The latter become attracted to the collector plates and are thus removed from the flue gas stream. Electrostatic precipitators will remove particles in the size range 0.005–1.0 microns ($10^{-6}$ metre), and are generally more than 99 per cent efficient at removing particulate material from the gas stream. The environmental fate of the small amount of solid material which escapes the precipitators is discussed below. It is the source of some of the radioactivity, heavy metals, and polycyclic aromatic hydrocarbon compounds emitted from power stations.

It is clear from the foregoing discussion that the main constituents of the exhaust stream are gases derived from combustion of the major elements in the fossil fuel, i.e. water vapour ($H_2O$), carbon dioxide ($CO_2$), sulphur dioxide ($SO_2$) and some of the nitric oxide (NO). The rest of the nitric oxide is produced, not from nitrogen in the fuel, but by oxidation of nitrogen ($N_2$) molecules in the air used in the burners. Other gases at low concentration in the exhaust include hydrogen chloride (HCl), nitrogen dioxide ($NO_2$), nitrous oxide ($N_2O$), carbon monoxide (CO) and sulphur trioxide ($SO_3$). In Table 2.4 the chemical composition of the emission from a typical coal-fired power plant is given.

A consideration of the data in Table 2.4 helps to identify the substances which should receive most attention here. The flue gas is, by virtue of the combustion process, depleted in oxygen and enriched in water vapour. However, since both of these gases are very abundant in the atmosphere, power station emissions have essentially no effect, except possibly very locally for water vapour. Examples of such local effects are the formation of visible plumes due to condensation of water vapour (from stacks and cooling towers) and occasionally in the production of convective clouds. The other major product of burning carbonaceous fuels is, of course, carbon dioxide ($CO_2$). Although this gas is the fifth most abundant component of air (after nitrogen, oxygen, water vapour, and argon) its concentration, at about 280 parts per million (p.p.m.) prior to industrialization, is low enough for emissions from fossil fuel burning to have had a measurable effect on its atmospheric level. Since $CO_2$ also plays a fundamental role in controlling the radiation balance and hence the climate of the earth, it will receive detailed treatment later in the chapter.

The other simple carbon-containing gas listed in the table is carbon monoxide (CO), which results from slight inefficiencies in the combustion

**Table 2.4** Chemical composition of stack emissions from a typical modern coal-fired power station (2000 MW)

(Concentrations in per cent (%), parts per million (p.p.m.), or parts per billion (p.p.b.) by volume, unless otherwise stated.)

| | |
|---|---|
| Air (oxygen depleted) | ~80% |
| $H_2O$ | ~4.5% |
| $CO_2$ | ~12% |
| CO | 40 (max. 1000) p.p.m. |
| $SO_2$ | 1000–1700 p.p.m. |
| $SO_3$ | 1–5 p.p.m. |
| NO | 400–600 p.p.m. |
| $NO_2$ | ~20 p.p.m. |
| $N_2O$ | ~40 p.p.m. |
| HCl | 250 p.p.m. |
| HF | <20 p.p.m. |
| Particulate material | <115 mg m$^{-3}$ |
| Hg (gaseous) | 3 p.p.b. |

*Source: Various CEGB publications*

process. Since concentrations of CO in industrialized areas are about 5 parts per billion (p.p.b.), power station emissions averaging about ten times this level have no significant effect within a very short distance of the power plant.

Next comes a group of various oxides of sulphur and nitrogen. Many of these compounds play important roles in determining the acidity of atmospheric aerosols and rain, certainly in industrialized regions, and will be considered in detail later. Within the sulphur oxides, sulphur dioxide ($SO_2$) is clearly more important than sulphur trioxide ($SO_3$), although in the atmosphere relatively slow conversion of $SO_2$ to $SO_3$ occurs. The major nitrogen oxide emitted is nitric oxide (NO), which is rather rapidly oxidized to nitrogen dioxide ($NO_2$) once it is in the atmosphere. In the emitted gas stream $NO_2$ and nitrous oxide ($N_2O$) are rather minor components.

Hydrochloric and hydrofluoric acids (HCl and HF) are not considered further here since they are not major components of the flue gas and their effects are generally to increase marginally the acidity largely provided by the sulphur and nitrogen oxides.

The final group of substances listed in Table 2.4 are mercury and particulate materials not removed by the precipitators.

The particulates contain polycyclic aromatic hydrocarbons, which are products of incomplete combustion, a variety of trace metals and radioactive isotopes. The figure of 115 mg m$^{-3}$ for particulates is an upper limit imposed on UK power stations by the Alkali Inspectorate (now H.M. Inspectorate of Pollution). None of these substances emitted at the allowed levels is thought to have any harmful effects on the environment. The possibility of their adversely affecting human health is returned to later in this chapter. The judgement on absence of environmental hazard is made both by comparing UK power station emission concentrations with emission standards set in countries where regulation is based on the use of such standards, and by seeing whether concentrations measured in the vicinity of power plants are significantly elevated above background levels or anywhere approach ambient air quality standards (CEGB 1980). Two interesting points emerge from such considerations. One is that although small amounts of radioactivity in particulate form are indeed released to the environment when coal is burned, this may be counterbalanced by a dilution of atmospheric $^{14}CO_2$ by gaseous $^{12}CO_2$. This comes about because the carbon in coal is very old (millions of years) relative to the half-life of $^{14}C$ (about 5700 years) and essentially 'dead' carbon is released to the atmosphere when fossil fuels are burnt. Another thing worth noting is that not all the emitted trace metals are in a particulate form. Mercury and to a lesser extent selenium are sufficiently volatile that in the combustion process they are released from the fuel in their elemental gaseous forms. Although some condensation on to fine particle surfaces occurs in the cooler parts of the system downstream of the burners, significant fractions (90% and 20% for Hg and Se respectively, (Klein *et al.* 1975)) of these metals are emitted from the power station chimney as gases. The dispersion pattern of mercury once in the atmosphere and before it has become involved in possible chemical reactions, will be like that of a gas rather than a substance in particulate form. This implies that it should be transported further and thus diluted more before eventual deposition on the ground.

## Carbon dioxide and the greenhouse effect

In order to assess the possible environmental effects of generating electricity by burning fossil fuels, we must examine what is known about temporal trends in atmospheric concentrations of carbon dioxide.

The earliest attempts to measure the concentration of carbon dioxide in the atmosphere were in the late nineteenth century. However, techniques of both sampling and analysis in use at that time meant that the values obtained were neither particularly accurate nor reproducible spatially or temporally. A selection of some of the better of the early direct measurements, as compiled by Callendar (1958), is shown in Fig. 2.2.

In fact, it was not until the late 1950s that reliable data became available. This came about almost entirely through the efforts of one man, Charles

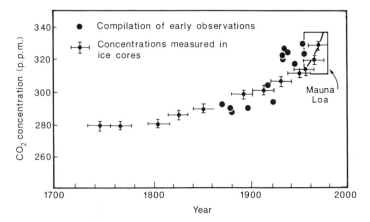

**Fig. 2.2** Atmospheric CO$_2$ concentrations over the period 1700–recent. ● compilation of early direct measurement data by Callendar (1958); ⊢●⊣ concentrations measured in ice cores by Neftel *et al* (1985); the bold line shows the recent data from Mauna Loa.

Keeling of The Scripps Institution of Oceanography in California. In 1958 he began a series of top quality measurements of atmospheric CO$_2$ at the Mauna Loa observatory on the Pacific island of Hawaii. The measurements continue to this day and constitute the 'bed-rock' of studies aimed at establishing man's influence on carbon dioxide in the atmosphere. The data from 1958–1980 are shown in Fig. 2.3. A clear increase in the yearly average values is apparent and by 1987 the level had reached 348 p.p.m. Although the amount of increase varies somewhat from year to year, it averages close to 1 p.p.m. per year for the period of the Mauna Loa record. Very similar increases are observed at other locations, although the records are generally not as long or as detailed as those from Hawaii. A selection of these other records, together with those from Mauna Loa are shown for the period 1958–74 in Fig. 2.4. It is plain from this figure that the increase in atmospheric CO$_2$ is a truly global phenomenon.

An interesting feature highlighted in Fig. 2.3 is the rather large regular seasonal changes that occur in carbon dioxide levels. This pattern is repeated at other observing stations. The amplitude of the seasonal cycle apparently shows a considerable variation with latitude, as illustrated in Fig. 2.5. These seasonal effects will be discussed further in connection with biological cycling of atmospheric CO$_2$ (pp. 31–2).

Although the observational record from Hawaii and other places gives a very good estimate of the rate of increase of CO$_2$ in the air since 1958, it would be of great interest to know values prior to this, particularly before the major industrialization which took place after the middle of the last century. As mentioned earlier, there are some direct measurements back into the late nineteenth century but they are of uncertain quality. Recently, another technique has become available for establishing historical changes in the chemical composition of the atmosphere. This involves the careful collection of ice cores from polar regions, and even more careful extraction and chemical analysis of the gases trapped at various depths in pores and interstices of the ice. The principle of the method is that the ice retains a record of the composition of the atmosphere at the time when the ice formed. By dating the various ice layers sampled, the time history of atmospheric chemical composition can potentially be established. Provided all the necessary precautions are taken, satisfactory results are obtained. This has been done for $CO_2$ on an ice core from West Antarctica by Neftel *et al.* (1985). The results are shown above in Fig. 2.2. and indicate a value of about 280 p.p.m. for the carbon dioxide level in the atmosphere in the eighteenth century, i.e. prior to the onset of major industrialization. Also shown on the Figure is the much shorter Mauna Loa record. The two

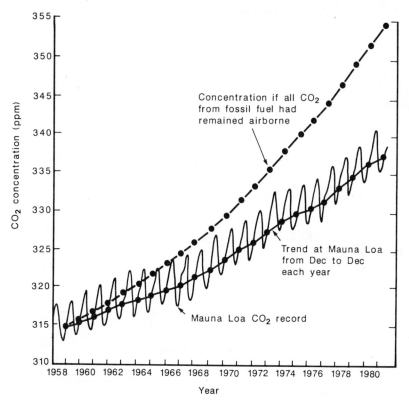

**Fig. 2.3** Carbon dioxide record from Mauna Loa, Hawaii, plus the predicted atmospheric concentration had all the $CO_2$ from fossil fuel burning remained airborne (from Crane and Liss 1985).

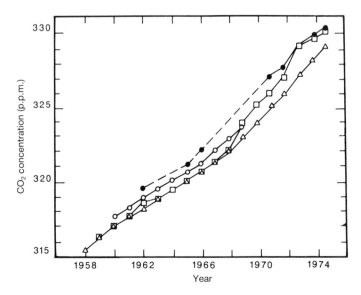

**Fig. 2.4** Yearly average atmospheric $CO_2$ concentrations over the period 1958–1976 at the following places : □ Mauna Loa; △ South Pole; ● Point Barrow, Alaska; ○ Swedish aircraft flights (N. Hemisphere at various heights) (after Kellogg 1978).

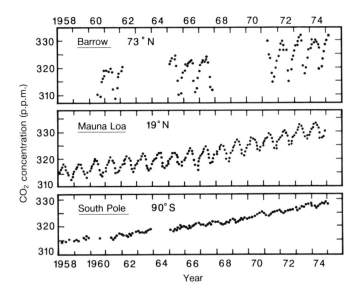

**Fig. 2.5** Monthly average atmospheric $CO_2$ concentrations over the period 1958–1975 at three locations (after Machta 1979).

data sets appear to agree well in terms of both general upward trend and overlap of the few common data points.

We thus have evidence for an increase in the level of carbon dioxide in the air over the last 200 years of about 25 per cent (from 280 to 350 p.p.m.). We now examine the various sources and sinks for atmospheric $CO_2$ with the aim of establishing how this increase has come about.

**Natural and anthropogenic sources (and sinks) of atmospheric carbon dioxide**

There are three major sources and sinks for atmospheric $CO_2$ in the environment; the land biosphere, the oceans, and man-made emissions from burning fossil fuels. As we shall see, the first two exchange $CO_2$ with the atmosphere in the natural state in an essentially balanced two-way transfer, but also play important roles as sources and sinks of anthropogenically produced carbon dioxide. Although volcanic emissions are sources of $CO_2$ to the atmosphere (and oceans), they are of minor importance on the time-scale being considered and will not be discussed further. The three major sources and sinks identified above will now be considered in turn.

*The terrestrial biosphere*

In its natural state the land surface exchanges approximately 100 GtC (gigatons, expressed as carbon; $1Gt = 10^9$ tonnes $= 10^{15}$ grams) each year with the atmosphere. This is a balanced steady-state flux, with 100 GtC going from land to air and the same amount going from the atmosphere to the ground each year. However, in temperate and polar regions the flux is seasonally unequal. Thus, in these regions in spring and summer, when plants are actively photosynthesizing and so abstracting $CO_2$ from the air, there is a net flux of $CO_2$ from the atmosphere to the land surface. On the other hand, in autumn and winter, when plant growth is small, the net flux is from the Earth's surface into the atmosphere owing to the dominance of the processes of respiration and decomposition of plant remains, which return $CO_2$ to the atmosphere. Averaged over the year there is no net flux either way. In tropical areas where there is little or no seasonality in plant growth the two-way fluxes occur simultaneously. Here, as in higher latitudes, there is always spatial variability in the up and down gross fluxes.

The seasonal asymmetry in the fluxes between land and air and vice versa described above is the reason for the seasonal cycle in the atmospheric $CO_2$ record discussed earlier (and shown in Figs 2.3 and 2.5). The falling values for atmospheric carbon dioxide in spring and summer are due to plant uptake removing the gas from the atmosphere, with the rising part of the yearly curve corresponding to net respiration and decomposition during the autumn and winter. The amplitude of the seasonal cyclicity

varies with latitude, as shown in Fig. 2.5. It is small (less than 1 p.p.m.) at
the South Pole owing to the absence of biological activity at this latitude.
At Mauna Loa (19° N) seasonal variation in plant growth leads to a signal
with an amplitude of greater than 3 p.p.m.—considerably greater than the
yearly average increase (about 1 p.p.m.). Further north at Barrow, Alaska
(73° N) the seasonal signal is even more pronounced, with an amplitude of
7–8 p.p.m. In general the amplitude of the 'saw-tooth' pattern in atmospheric $CO_2$ is greater in the northern compared to the southern hemisphere
owing to the considerably greater land area in the former, relative to the
latter.

We can conclude that although anthropogenic activities appear to control
the year-by-year increase, biological processes seem to dominate the seasonal
pattern of atmospheric carbon dioxide. This clearly demonstrates the importance of biological processes on land in affecting levels of $CO_2$ and
raises the interesting question of whether alteration of land use by man
could have resulted in net transfers of carbon between land and air. It has
often been argued that by converting virgin forest and other land areas
having large amounts of 'fixed' (i.e. plant) carbon to agricultural, urban,
and industrial use, where the stock of fixed carbon is much less, net release
of carbon dioxide must have occurred. However, it has proved very
difficult to arrive at a generally accepted figure for this potential net flux to
the atmosphere. Some of the more reliable estimates for the period 1860–
1980 are shown in Fig. 2.6. Bolin (1986) has concluded that the best

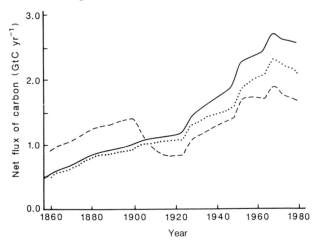

**Fig. 2.6** Emissions of $CO_2$ to the atmosphere resulting from deforestation and
changing land use, according to three different procedures for analysing past data
(from Houghton 1984, and Bolin *et al* 1985). The solid line is based on biomass data
from Whittaker and Likens (1975); the dotted line is based on tropical biomass
derived from timber volumes reported by FAO; the dashed line depends on
another scenario for the transfer of forest land and grassland into agriculture.

estimate of the net flux of carbon to the atmosphere as a result of change in land use in 1980 is $1.6 \pm 0.8$ GtC per year. Against this must be set the possibility that the higher levels of atmospheric $CO_2$ pertaining now may have led to greater rates of plant photosynthesis, and hence fixation of carbon, than previously. Although there is evidence for this from crops grown under glass with artificially elevated $CO_2$ levels, it seems less likely to apply in the world at large, where such factors as availability of nutrient and water are generally those that limit plant growth. However, there is currently insufficient evidence to dismiss the idea out of hand. All we can safely say at present is that the net input to the atmosphere of carbon dioxide from changes in land use is probably somewhere in the range 0–2 GtC per year.

*The oceans*

As with the land biosphere, the oceans also exchange large amounts of carbon dioxide with the unpolluted atmosphere each year in a balanced steady-state flux. The magnitude of these two-way fluxes is again approximately 100 GtC per year in each direction. The fluxes are in part driven by temperature changes in the surface oceans altering the solubility of $CO_2$ in the water, and partly by consumption and production of the gas by photosynthetic marine plants and respiration/decomposition processes in surface waters. All of these processes are both spatially and seasonally variable. It is possible to say that in very general terms tropical areas are net sources of $CO_2$ to the atmosphere, whereas at higher latitudes, particularly in polar regions, the oceans act as a net sink. Further, taken as a whole and averaging over a yearly period, the ocean–atmosphere system in its unpolluted state approximates to a steady-state condition. This does not mean that the system does not alter over long periods. For example, the substantially lower atmospheric carbon dioxide levels which seem to have pertained in glacial periods (down to 200 p.p.m. during the last glaciation, according to analyses of air trapped in ice cores) are generally attributed to enhanced ocean uptake then as compared with the present.

The above discussion has been about atmosphere–ocean exchange of $CO_2$ before the major industrial use of fossil fuels. As we shall quantify in the next section, and is apparent from the recent atmospheric record described previously, man has injected large amounts of extra carbon dioxide into the atmosphere. How much of this extra $CO_2$ has been taken up by the oceans as a net air-to-sea transfer?

In trying to answer this important question quantitatively, several factors have to be taken into account. One is the chemistry of sea-water itself. Relative to distilled water or a solution of sodium chloride of similar salt content to ocean waters, sea-water is a better absorber of excess carbon dioxide. This arises from the presence of alkalinity in the form of carbonate ions in the sea-water. The carbonate ions react with absorbed $CO_2$ to form bicarbonate ions, so that effectively dissolution is enhanced by reaction with the dissolved carbonate ions. To get an idea of the magnitude of this effect, it can be calculated that sea-water is about eight times more able to take up extra $CO_2$ than a similar water but with the alkalinity-producing carbonate ions replaced by ions ineffective at reacting with dissolved carbon dioxide (e.g. chloride, sulphate).

This calculation of the ability of sea-water to take up excess atmospheric carbon dioxide assumes that equilibrium is achieved between the air and the water. With a body of fluid of the size and particularly the depth of the oceans this assumption is unlikely to be true, since it would take hundreds if not thousands of years for the oceans to become fully equilibriated to a new atmospheric concentration of $CO_2$. The major impediment to the attainment of equilibrium is not the rate of transfer of carbon dioxide across the sea surface (although this can be a factor in more sophisticated calculations), but the rate at which surface sea-water enriched in $CO_2$ can move down into the deeper parts of the oceans. Such movement is impeded greatly by the existence in most areas of the oceans of a stable two-layer structure in the water column. At a depth of several hundred metres, called the main thermocline, there is a rapid decrease in temperature and hence increase in water stability which inhibits vertical mixing of the water from above and below. It is only in polar regions, particularly around Antarctica and in the Greenland and Norwegian Seas of the North Atlantic, that direct communication between the surface and deep water occurs.

In order to quantify the amount of anthropogenic carbon dioxide the oceans can take up from the atmosphere it is necessary to construct models of the system. These often take the form of a set of boxes (numbering from a few to several hundred) with water containing its dissolved carbon flowing between them. The main elements of many of the models are illustrated in Fig. 2.7. For the well-mixed atmospheric and surface ocean boxes the flow between them is assumed to be proportional to their carbon content. For the deeper boxes, flow within and between them is often modelled as a diffusion process, and direct flows from surface to deep can also be incorporated. Radioactive substances such as carbon-14 (produced both by cosmic rays in the atmosphere and from the detonation of nuclear devices) can then be used to obtain the coefficients describing the transport rates of the various flows. Using such models the fraction of a given increment of $CO_2$ added to the atmosphere that will be taken up by the ocean reservoir can be calculated. These models suggest that about 35 per cent of anthropogenic carbon dioxide from fossil fuel burning is absorbed by the oceans. This corresponds to roughly 2 GtC per year for present rates of fossil fuel combustion.

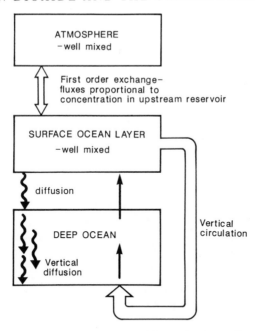

**Fig. 2.7**   A simple model for the uptake of $CO_2$ by the ocean (from Crane and Liss 1985).

### Fossil fuel combustion

In a way, input of carbon dioxide to the atmosphere as a result of burning fossil fuels is the easiest major source to deal with. This is because there is no natural component to be considered and this certainly simplifies quantification of the size of the input. All that is needed is the amount of each particular type of fuel consumed in the period under consideration and a knowledge of the quantity of $CO_2$ produced from the burning of a unit amount of each sort of fuel. Such data are available from the annual United Nations *Energy Statistics Yearbook*. Recently the data have been corrected for flaring of natural gas and for the small contribution of $CO_2$ to the atmosphere which arises during the manufacture of cement.

The results of such studies are shown in Fig. 2.8, which clearly highlights the expected rapid rise in emissions with time, and particularly during the last forty years. Further information can be extracted from the data if they are plotted on a logarithmic scale, as shown in Fig. 2.9. With well-defined exceptions, it is apparent that the *rate* of increase in emissions has been remarkably constant up until the early 1970s at close to 4.3 per cent per year. The exceptions are the 'Great Depression' of the 1930s and the two world wars, when the rate of increase considerably slackened. The sharp

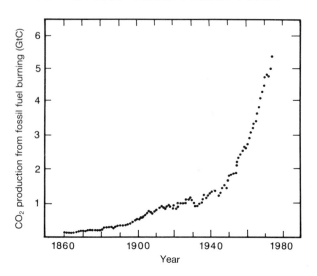

**Fig. 2.8**   Annual production of $CO_2$ from fossil fuels and cement 1860–1977 (from Rotty 1980).

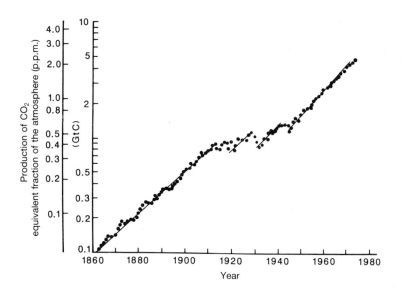

**Fig. 2.9**   Annual production of $CO_2$ from fossil fuels and cement 1860–1974—logarithmic plot (from Rotty 1977).

increase in the cost of fossil fuels which occurred in the 1970s led to another slowing in the growth rate (as illustrated in Fig. 2.10), owing in part to more efficient use of energy. Although the overall rate of increase in emissions from about 1972 to 1980 was halved from its 'historical' value to about 2.25 per cent, the effect was uneven between different types of fuels. Coal showed essentially no change from its previous rate of increase of just under 2 per cent, whereas there was a pronounced slowing down in the rate of increase for gas and particularly for oil, which had been rising at 7–8 per cent prior to this time (Fig. 2.11).

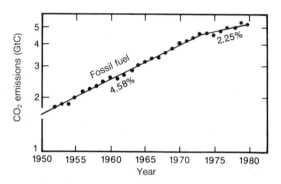

**Fig. 2.10**   Annual global production of $CO_2$ from fossil fuels and cement—logarithmic plot, with indicated growth rate (from Rotty 1983).

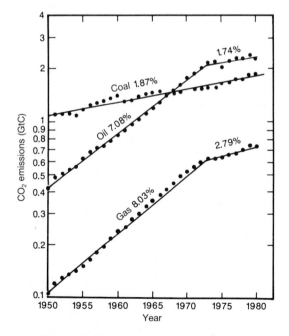

**Fig. 2.11**   Annual $CO_2$ emissions from each fuel type—logarithmic plot, with indicated growth rates (from Rotty 1983).

At the present time the emission of $CO_2$ to the atmosphere from fossil fuel consumption is about 5 GtC per year. Of this approximately 30 per cent comes from conventional power stations (this is close to the figure for the US and the UK, and may be somewhat on the high side for less-developed countries).

**The global budget of man-made carbon dioxide**

Much of what was discussed in the previous section on sources of atmospheric $CO_2$ is summarized and put into a global context in Fig. 2.12. The first thing to notice on the diagram is the very different sizes of the various reservoirs of carbon. Although the sediments are by far the biggest reservoir, mainly in various forms of calcium carbonate, most of this material is not in contact with the atmosphere, at least on the relatively short time-scale of interest here. The next biggest reservoir is the oceans where the carbon is mostly in the form of bicarbonate and carbonate ions. However, as we saw previously, the deep layer, which constitutes by far the greater part, does not interact with the atmosphere at all quickly. The reservoir of carbon in fossil fuel and shales is also large and a major fraction of it is thought to be ultimately recoverable for use as fossil fuel, should man choose to do so. The smallest reservoirs are the land biosphere

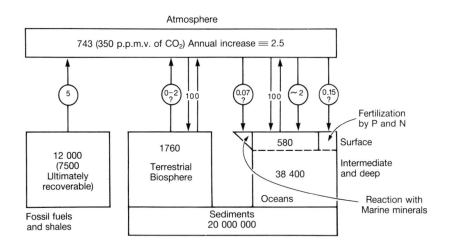

**Fig. 2.12** Global carbon reservoirs and present natural and anthropogenic fluxes between reservoirs. Reservoir sizes in GtC. Fluxes between reservoirs in GtC per year. Anthropogenic fluxes are circled. (From Liss and Crane 1983).

and the atmosphere. It is the small size of the latter which makes it susceptible to significant perturbation by even small changes in the flows in or out of the other reservoirs (e.g. mobilization of the fossil fuel reservoir by burning).

This leads us to the flows of carbon between the reservoirs. The previously discussed steady-state fluxes of approximately 100 GtC per year each between the atmosphere and the land biosphere and the oceans, are also shown in Fig. 2.12. The net input to the atmosphere of about 5 GtC per year from fossil fuel burning is also shown. Of this close to half (2.5 GtC per year) appears to remain in the atmosphere and leads to the rising trend in air concentrations of $CO_2$ apparent in Fig. 2.3. Also shown in Fig. 2.3 is what the atmospheric concentration would have looked like if all the carbon dioxide emitted from fossil fuel burning since 1958 had stayed in the atmosphere. Clearly, about half of it does not appear in the atmospheric record and must have gone elsewhere. The obvious sink is the oceans which, as already discussed, appear able to accept about 2 GtC per year. Thus, there would seem to be an imbalance of approximately 0.5 GtC per year, at a minimum. Any net emission as result of man-induced changes in terrestrial biosphere will serve to exacerbate the discrepancy, which could then be as high as 2.5 GtC per year if the maximum value for net anthropogenic biospheric emission given in Fig. 2.12 were proved to be correct.

This imbalance in the budgeting of man-made carbon dioxide has led people to investigate whether any other natural processes exist for removing $CO_2$ from the atmosphere. Enhanced growth of plants due to elevated carbon dioxide levels has already been discussed and seems unlikely to do more than back-off some of the net input due to land-use change, but even this small degree of offset is far from certain.

Another suggestion is that since rivers have become enriched in plant nutrients (principally nitrate and other forms of nitrogen and phosphate) owing to agricultural and urban development, input of these waters into the marine environment may have led to increased plankton growth, hence carbon fixation and ultimately sedimentation, in the oceans. Although the effect is probably real, estimates of its magnitude indicate that it could lead to sequestration of only relatively small amounts of carbon dioxide. A figure of 0.15 GtC per year is shown in Fig. 2.12.

A third possibility is that excess $CO_2$ taken up by the oceans reacts with mineral phases in the marine environment. Heterogeneous reactions such as this and the biological uptake mechanism just discussed are explicitly excluded from ocean uptake models which are presently formulated solely in terms of homogeneous processes. The $CO_2$–mineral interaction mechanism is potentially very powerful in view of the large mass of calcareous sediments on large areas of the sea floor. However, much of this is clearly not in contact with surface waters, so the time-scale of any reaction is likely to be long. Further, near-surface sea-water appears to be supersaturated with respect to the pure calcium carbonate mineral phases calcite and aragonite, so that precipitation rather than dissolution is the favoured process. The only solid phases likely to dissolve are calcium carbonate minerals containing significant amounts of magnesium, which makes them considerably more soluble. The extent to which such minerals do indeed dissolve in response to additional $CO_2$ entering the oceans is difficult to estimate but the best figures, for example the value of 0.07 GtC per year given in Fig. 2.12, imply that this is also a relatively small sink.

In conclusion, it seems that the global budget for anthropogenically produced atmospheric carbon dioxide is not well understood quantitatively. The apparent imbalance implies that we are either ignorant of additional mechanisms for removing carbon dioxide from the atmosphere and/or our quantitative appreciation of the identified processes is far from satisfactory. Until this problem is resolved, predictions of the proportion of future emissions of $CO_2$ which will remain in the atmosphere will be subject to considerable uncertainty.

**The potential climatic role of man-made atmospheric carbon dioxide and resulting environmental impacts**

Although only a relatively minor component of the atmosphere, carbon dioxide plays a vital role in the radiation balance of the earth and hence in controlling its climate. This is best illustrated by reference to Fig. 2.13. The upper part of the diagram (a) shows the wavelength emission spectrum of the sun and the earth, at their effective radiating temperatures of about 5700°C and −23°C, respectively. The lower half (b) illustrates how this emitted radiation is absorbed by various gaseous components of the atmosphere. For example, much of the ultra-violet radiation arriving at the earth from the sun is absorbed by ozone molecules high in the atmosphere and explains the current concern that man-induced decreases in stratospheric ozone may lead to larger amounts of harmful ultra-violet radiation reaching the surface of the earth. Most of the rest of the solar radiation reaching the earth passes through the atmosphere to the ground without major absorption. Of great relevance here is the absorption band near the middle of the earth's emission spectrum (centred around 15 $\mu$m) resulting from the presence of carbon dioxide molecules. This, together with other absorption bands composed of water molecules, means that the atmosphere and surface earth are considerably warmer (global mean surface temperature close to 15°C) than the effective emission temperature of the earth (−23°C).

The combined effect of the atmosphere's transparency to most of the incoming solar radiation and the absorption of much of the earth's emitted radiation by water vapour and $CO_2$ in the atmosphere explain why these

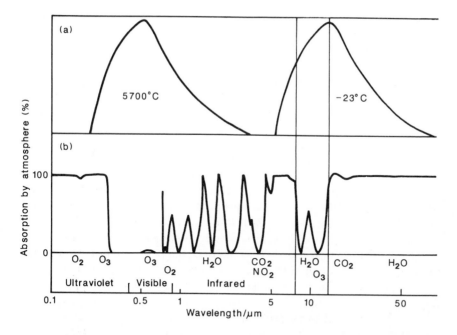

**Fig. 2.13**  a) Radiation spectra for the sun (5700°C) and (not to same scale) the earth (−23°C). b) Atmospheric absorption spectrum produced by the principal absorbing atmospheric gases. (From Spedding 1974).

combined properties are often referred to as the 'greenhouse effect'. In a greenhouse the glass serves to let the sunlight through but prevents the escape of much of the radiant energy coming from the soil and other surfaces within. The analogy is only approximate since, for example, the transmission/absorption properties of the glass do not precisely match those of $H_2O$ and $CO_2$ molecules, and in a greenhouse much of the warming is due to prevention of convective heat loss to the free atmosphere by the structure itself.

From the reasoning given so far in this section it is easy to see why elevated levels of atmospheric carbon dioxide resulting from man's activities may lead to an increase in the temperature of the atmosphere. Put simply, if the temperature of the pristine atmosphere is higher than it would otherwise be because of the presence of carbon dioxide, then additional $CO_2$ from burning or any other source will increase the temperature still further. This simple line of reasoning, however, tells only part of the story. Inspection of Fig. 2.13 reveals that unpolluted levels of $CO_2$ are sufficient to absorb almost 100 per cent of energy coming from the earth in the relevant wavelength bands. Although the $CO_2$ absorption bands will broaden somewhat with elevated concentrations of the gas in the air, the major effect is a vertical redistribution of the heights at which the absorption occurs. With more carbon dioxide molecules in the air there will be greater absorption at lower levels in the atmosphere and correspondingly less at higher altitudes. The result of this is that the layers of the atmosphere nearer to the ground will warm, while higher levels will concomitantly cool. As we shall see shortly, these rather qualitative predictions are indeed confirmed by sophisticated modelling approaches.

Mathematical models of various degrees of complexity have been devised in order to try to predict quantitatively the warming of the troposphere (and cooling of the stratosphere) which are likely to occur as the atmospheric carbon dioxide level continues to rise in the future as more and more fossil fuel is burned (and forest converted to other purposes). Most of the models work on the premise that at some future date the atmospheric concentration of carbon dioxide will reach double its pre-industrial value (i.e. it will rise from about 280 to 560 p.p.m.). As we shall see in the next section, this is very likely to happen sometime during the next century, although there are too many uncertainties for an exact estimate of the timing to be made. Notwithstanding the uncertainties, the doubling scenario is arguably a relevant one to take since there are people alive today who will in all probability live to see it happen. Further, the temperature changes to be expected from a doubling of $CO_2$ turn out to be in the range of such alterations which have occurred in the past (e.g. between glacial and inter-glacial periods) and so are environmentally realistic.

The simplest models assume that the atmosphere is everywhere at radiative equilibrium (incoming and outgoing thermal energy are equal) and examine what a doubling of carbon dioxide in the atmosphere will do to the surface temperature. The result turns out to be a warming of about 2.8°C. More sophisticated models allow atmospheric motions to occur and are thus considerably more realistic. However, computational and other constraints lead to simplifications in the number of length dimensions (height, latitude, and longitude) which are explicitly included. Some models have the atmosphere as horizontally homogeneous and only vertical motion is included (called one-dimensional models). Somewhat more complete representations have both vertical and latitudinal variations (two-dimensional models). The most complete models include all three dimensions and are often called general circulation models (GCMs). The outcome from all of these models is generally a mean warming of between 2 and 4°C, in agreement with the simplest equilibrium approaches. Although the agreement in the average temperature change is good, the models, of course, differ in the degree of horizontal and vertical detail they can predict.

This is well illustrated in Fig. 2.14, which shows the zonally averaged output from a typical GCM. In this model the average temperature rise

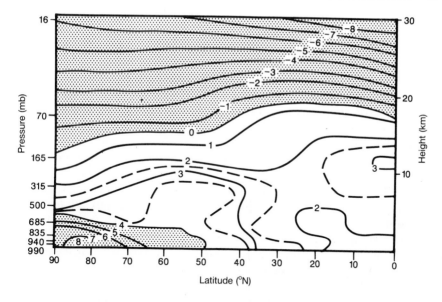

**Fig. 2.14**  Latitude–height distribution in the change in zonal-mean temperature (°C) in response to a doubling in atmospheric $CO_2$ (from Manabe and Wetherald 1980).

near the ground is 3°C, but spatially the temperatures show marked variations. They are less than average at low latitude but considerably more north of 50°. The greater warming at high latitudes is due to several factors, including decreased ice cover leading to increased absorption of radiation by the ground, and stable conditions inhibiting vertical mixing so that the $CO_2$-induced warming is confined to the lowest atmospheric layers. As predicted earlier, the stratosphere cools as the troposphere warms.

So far we have concentrated only on $CO_2$-induced temperature changes as predicted by general circulation models. However, other meterological changes, e.g. in precipitation, may be more important in a practical sense than temperature alteration *per se*. The major results from GCM runs indicate that in middle and high latitudes summers are likely to be drier and in high latitudes winters wetter under a doubled $CO_2$ regime. Alterations will also occur in evaporation rates; the change in the difference between precipitation and evaporation amounts determining how the soil moisture content will be affected. Increasing summer dryness in the largely mid-latitude major grain-growing areas of the world will, if the model predictions prove accurate, produce obvious deleterious effects on crop yields which will have considerable social, economic and political consequences.

However, the predictions of even the most sophisticated general circulation models have very real uncertainties attached to them. In general, these arise either from simplifications which have to be made in order to run the model with the size of computer available, or because we do not yet know how to specify changes which will occur in some of the basic elements of the system as $CO_2$ increases and the climate warms. Examples of simplifications that have been adopted for computational purposes include idealized geography and omission of seasonal patterns, particularly if fine spatial resolution is required. Further, most GCMs currently in use compute the difference between present (or pre-industrial) conditions and those which will result from a doubling of $CO_2$. Instead of such a step function approach, it is more realistic to model what happens with a gradually rising level of atmospheric $CO_2$, as recently attempted by Hansen *et al.* (1988).

The two most difficult factors to incorporate properly into general circulation and other models are, given a $CO_2$-induced warming, what effect does this have on the extent and nature of cloud cover and how do the oceans respond to the warming of the atmosphere?

Since clouds have too small a vertical or horizontal extent to be modelled explicitly in global-scale grid models, the early climate models assumed no change in cloud cover with doubled $CO_2$ levels. This is unlikely to be correct. It can be argued that increased air temperatures will lead to greater evaporation and hence more cloudiness. Clouds certainly play important roles in controlling the amount of solar radiation reaching the earth's surface; it has been calculated that an increase of only 1 per cent in cloud cover could counteract the effect in terms of temperature of a 25 per cent addition to atmospheric carbon dioxide. However, other workers have argued that increased evaporation will lead to a change in cloud type (more convective, less stratiform) with the net effect being a decrease in cloud cover. Most of the more recent models do incorporate cloud parametrization schemes which allow cloud characteristics to change as climate warming proceeds. Nevertheless, they do not simulate all the factors regarded as important for calculating the radiative changes, and no one would yet regard the modelling of clouds as anywhere near satisfactory.

As for the response of the ocean surface waters to increased air temperatures, one of two opposed and rather extreme assumptions are often made in general circulation and other models. In the models the ocean supplies water vapour to the atmosphere. Where the models differ is in the way the thermal capacity of the surface water is treated. In some, the water temperature is held at the climatologically observed temperatures, which effectively means the ocean is assumed to have an infinite thermal capacity and change in air temperature produces no alteration in the temperature of the sea-water. The alternative assumption is that the ocean has no heat capacity and just adopts the temperature of the air, i.e. the ocean acts merely as a wet or 'swamp' surface. This latter is the more generally used assumption. Which assumption is used makes a significant difference to the final predictions of the amount of atmospheric warming. For example, in GCM simulations using the above two different assumptions it was found that the predicted warming for a doubling of atmospheric $CO_2$ was 2–3°C for the swamp ocean and only 0.2°C using the climatological water temperatures. The reason for this rather big discrepancy is that with the climatological water temperatures there is only a small increase in atmospheric water vapour as the air warms. This means that the warming due to $CO_2$ increase alone would be quite small, with most of the greenhouse effect coming from the increasing amounts of water vapour entering the atmosphere as the air and swamp ocean surface increase in temperature. Although the above example is probably a severe test of the models, it does illustrate their sensitivity to the assumptions used with regard to the thermal capacity of the surface oceans. Recently this has become less of a problem with the advent of coupled atmosphere–ocean models which, although computationally much more demanding, do incorporate air–sea interactions in a physically more realistic manner.

Mention should be made of two further oceanographic factors which may be important. One is that in some climate models transport of heat horizontally by ocean currents is largely ignored. This is potentially important since such transport is believed to carry up to a half of the total heat flowing polewards in the present atmosphere/ocean system. A second is that recently a proposal has been made by Charlson *et al.* (1987) that atmospheric sulphate aerosols, formed by atmospheric oxidation of a gas called dimethyl sulphide produced by plankton living in surface sea-water, may function as a climate regulating system which acts in the opposite sense to that of carbon dioxide. We will return to this intriguing possibility in a later section where natural sources of atmospheric sulphur are discussed.

Finally we should note that the effect of increasing atmospheric carbon dioxide is not solely to raise the temperature of the troposphere and surface of the earth. Many other meteorological, hydrological, and oceanographic parameters are also affected; some have been referred to already. Another important outcome of a warmer earth would be a rise in sea level. This would come about in part from thermal expansion of sea-water and also as a result of the melting of glaciers and small ice caps. Although there are considerable uncertainties in the predictions, best estimates of the combined effect for a global warming of about 3°C are for a rise in sea level of the order of 1 metre. If it occurred this would have very significant effects on the many countries that have large centres of population close to the sea or on low-lying land. Further, there is a possibility that the warming might eventually lead to the melting of a large mass of grounded ice, for example the West Antarctic ice sheet. Such an event could produce a rise in sea level of about 5 metres but, even if 3°C is a sufficient warming to melt the ice, it would probably take several hundred years for this to occur.

**Possible implications of carbon dioxide-induced warming for power generation**

In this section we consider what implications the predicted climate warming arising from increased atmospheric carbon dioxide may have on power generation in the future. Although power stations currently account for only about 30 per cent of the $CO_2$ entering the atmosphere from fossil fuel combustion, they are large, fixed, readily identifiable units. As such they may be perceived as more amenable to regulation than the whole host of small, often mobile, sources which presently burn the majority of the fossil fuel. It thus seems sensible to consider the $CO_2$-climate issue as it may affect power stations in the future. This will be done by considering the following four questions:

1. What is the timescale of future atmospheric concentrations of $CO_2$?

2. When will any warming be observable?

3. What is the likely effect of climatic warming on demand for power?

4. What action can be taken to delay or reduce $CO_2$-induced climatic change?

*What is the timescale of future atmospheric carbon dioxide concentrations?*

In order to predict future air concentrations of $CO_2$ we start with the well-established observations of current levels from Mauna Loa or elsewhere. Then several assumptions have to be made, including the mix of fossil fuels which will be burned in the future (generally taken to be as now); the partitioning of the emitted gas between the atmosphere and other reservoirs, principally the oceans; and the rate of increase of fossil fuel burning over the time period considered.

The assumption of continuation of current fuel mix is probably conservative, in that any shift from gas and oil towards coal or synthetic fuels made from coal will lead to increased $CO_2$ emissions for the same amount of energy produced, as illustrated in Table 2.5. Such a shift is probably inevitable in the longer term in view of the much larger reserves of coal compared to oil and gas. However, the timing and extent of the switch to coal is difficult to assess and the assumption of continuation of the present mix is probably the best that can be done.

Predicting the proportion of emitted $CO_2$ which will stay in the atmosphere is similarly a problem, particularly in view of our lack of

**Table 2.5**   Carbon dioxide produced per unit of energy generated from different fuel types

| Fuel | Carbon (as $CO_2$) produced per 100 Quads* of thermal energy (GtC) |
|---|---|
| Oil | 2.0 |
| Gas | 1.45 |
| Coal | 2.5 |
| Synfuel (derived from coal) | 3.4 |

*1 Quad $=10^{15}$ BTU $(=1.055 \times 10^{15}$ kJ$)$

knowledge of even the present global budgeting of man-made carbon dioxide, as discussed in an earlier section. What is done is to assume a central value which is calculated as the historical ratio of observed atmospheric $CO_2$ increase at Mauna Loa to the amount emitted by fossil fuel consumption during the same time period (the so-called 'airborne fraction' ($\alpha$), and equal to 0.55). To try to deal with the uncertainty as to whether $\alpha=0.55$ will apply in the future, the calculation is also performed using high ($\alpha=0.67$) and low ($\alpha=0.38$) values of the airborne fraction. The calculation of $\alpha=0.55$ assumes that the biosphere has been neutral, i.e. neither a net source nor sink. The extreme values encompass the range of possible airborne fractions in the past with a biosphere whose size has changed with time.

The most difficult and, as it turns out, the most important assumption is what the rate of increase in fossil fuel usage will be in the future. Over the last century the rate has been about 4.5 per cent per year, but has slackened to just over 2 per cent per year since the early 1970s (as illustrated in Fig. 2.10). Many guesses of what this rate of increase will be in the future can be made. Here we choose to repeat the calculation using three different rates; 4 per cent per year, representative of the long-term historical rate; 2 per cent per year, which is close to the post 1973 rate; and zero growth in fossil fuel burning. Although it is probably unlikely that the rate will remain constant, it seems likely that it will at least remain within these bounds. It is improbable that a growth rate of more than 4 per cent per annum could be sustained and zero growth looks very unlikely, particularly in view of the aspirations of countries in the developing world (and see Fig. 2.16).

The results of these projections are shown in Fig. 2.15. It is immediately apparent from the figure that the range of values used for the airborne fraction make only small differences to the results. On the other hand, the value adopted for the growth rate has a very large effect on the outcome. If

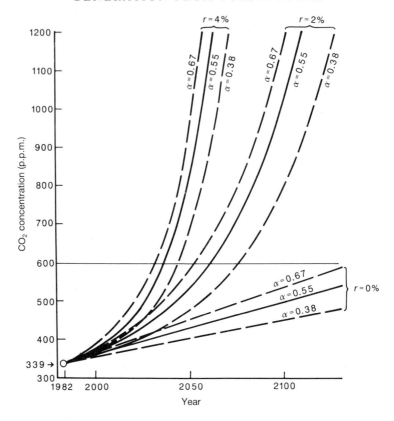

**Fig. 2.15** Predicted growth in atmospheric $CO_2$ concentration over the next 150 years, assuming growth rates ($r$) in $CO_2$ emissions of 0, 2 and 4% per year, and airborne fractions ($\alpha$) of 0.38, 0.55, and 0.67 (from Liss and Crane 1983).

an arbitrary atmospheric $CO_2$ concentration of 600 p.p.m. is chosen (this is indeed a value used in many climate prediction models since it is approximately double the pre-industrial level) then with a growth rate of 4 per cent per year this level will be achieved early in the next century (central value 2036, with a range from 2032 to 2045 for $\alpha$ at 0.67 and 0.38, respectively). With a growth rate of 2 per cent per annum 600 p.p.m. is reached in 2060 ($\alpha=0.55$) with a range from 2053–2076. With zero growth the doubling is delayed well into the twenty-second century (2172, with a range from 2138 to 2258). It is clear that unless the rate of growth in fossil fuel consumption is kept well below 2 per cent per annum, atmospheric carbon dioxide will be double its pre-industrial level sometime in the next century.

It is evident from the above that growth in fossil fuel usage is the single most important factor controlling atmospheric $CO_2$ in the future. Assessments of where geographically this growth is likely to occur have been attempted. One of these is shown in Fig. 2.16 for the period from 1974 to 2025. From this it is apparent that growth in the currently developed nations will be much less than that predicted for the presently less developed countries of Asia, Africa and South America.

It is worth noting in the context of the timing of atmospheric $CO_2$ increase that carbon dioxide and water vapour are not the only 'greenhouse' gases. Several other gases produced by anthropogenic activities are radiatively active in the atmosphere. These include ozone, methane, chlorofluorcarbons, and nitrous oxide. Their sources are various and the direct contribution from power stations is almost certainly small. However, these gases are likely to become important in the future if their current rising trends in the atmosphere continue (emission of chlorofluorocarbons is now actively being regulated by international agreement). Although difficult to predict with any certainty at the present, the overall effect of these other gases must be to advance the timing of any greenhouse warming.

*When will any warming be observable?*

In the previous section we examined when atmospheric carbon dioxide levels might double. The climatic models certainly indicate that such an increase will lead to a large change in climate (one would have to go back more than 10 000 years in order to encounter a natural climatic change of this magnitude, but of opposite sign). However, from a practical point of view smaller degrees of climatic change will almost certainly have significant, and possibly disastrous, effects on human society. It has been argued that nothing will (or, possibly, should) be done in earnest to try to forestall the rising trend of atmospheric carbon dioxide until a warming is definitely established. This highlights the importance of the question posed here, i.e. when will it be possible to detect with statistical certainty a $CO_2$-induced warming against the natural variability of our climate?

Several researchers have tried to address this difficult question. Most conclude that the global warming should be apparent some time in the 1990s. The results of one of these studies are given in Fig. 2.17. It shows the historical temperature record (plotted as deviations from the mean) since 1950 and a one-dimensional model is used to predict this line into the future. The prediction is then superimposed on plots of one and two standard deviations of the observed natural variability of temperature (again expressed as deviations from the mean) projected forward in time to 2020. The predicted temperature deviation line crosses the one-sigma (85 per cent confidence limit) and the two-sigma (98 per cent confidence limit) boundaries in the mid-1980s and mid-1990s, respectively. Although predictions of the sort shown in Fig. 2.17 were treated with caution by many scientists when they were first published in the early 1980s, there now seems to be a growing body of opinion that an observable $CO_2$-induced

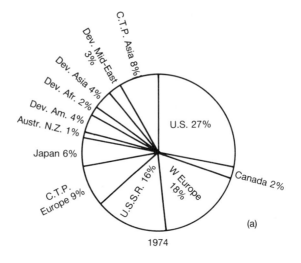

1974 (a)

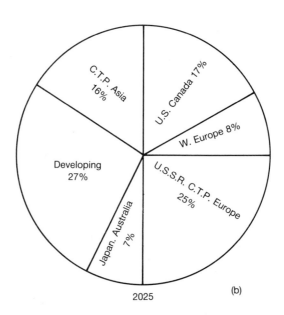

2025 (b)

**Fig. 2.16** Global $CO_2$ production by world segments (a) for 1974 and (b) estimated for 2025. (C.T.P. = centrally planned economies). Disc area in proportion to total production. After Marland and Rotty (1980).

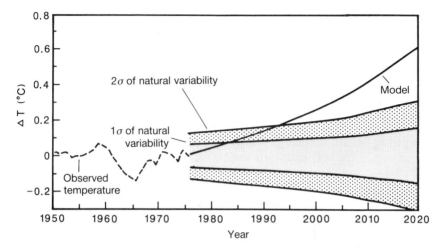

**Fig. 2.17** Carbon dioxide warming versus natural climate variability. Comparison of $CO_2$-induced warming predicted from a climate model with one- and two-standard deviations ($\sigma$) of the observed global temperature record. (After Hansen *et al* 1981).

warming is either already here or will be in the very near future. This is partly because later and more sophisticated climate models have tended to confirm the predictions of the earlier simpler models. Another factor is that more detailed analysis of the record of global temperatures, which gets both longer (inevitably) and more reliable as time goes on, indicates a significant warming in recent years in line with the model predictions. Other factors which tend to get a great deal of publicity are the finding that the four warmest years in the past century were all in the 1980s, and the heat wave and drought in the US Midwest at the time when this is being written (summer 1988) with the consequences of poor grain yields in that globally important agricultural region. Such important and well publicized impacts of climate on agriculture, coupled with the firming-up of the scientific consensus on global warming, probably go a long way to account for the recent increased governmental interest in controlling $CO_2$ emissions. An example is the intergovernmental conference held in Toronto, Canada in June 1988 where, for example, it was agreed that the wealthy industrialized nations should aim to decrease their $CO_2$ emissions by 20 per cent of 1988 levels by the year 2005.

### What is the likely effect of climatic warming on demand for power?

With a warmer climate the actual demand for electrical power is likely to be altered from the present situation. It is difficult to deal with this question in global terms since the use to which power is put, and hence the temperature-sensitivity of the demand, differs widely between countries, particularly depending on the local climate. With respect to the UK some idea can be gained from looking at the present way demand for electric power alters as temperature changes daily and seasonally.

Averaged over a year in the UK the demand for electricity varies inversely by approximately 1% per 1°C change in temperature. That is, for a temperature increase of 1°C above average the demand for electricity drops by about 1 per cent of that normally used. This is a rather small effect compared with other energy supply industries (e.g. 4–5 per cent per °C for oil and gas), owing mainly to the fact that electricity in Britain is used relatively little for space heating, where temperature has the greatest effect on demand. Even in the spring and autumn when the temperature sensitivity of electricity supply is greatest, the effect is still only about 2% per °C. Applying these sensitivities in the future to a climate significantly warmer than at present is plainly a long extrapolation. It assumes that no changes in building technology and practice will take place which will reduce the above quoted sensitivities. Further it could well be that our response to a generally warmer climate will be to install more air conditioning plant in buildings. If, as at present, these are powered by electricity this would not only decrease the overall temperature sensitivity but also reduce the current winter/summer differential in electricity usage.

The above analysis shows that for a climatic warming of say, 3°C the demand for electricity will decrease by approximately 3 per cent on average (perhaps 5 per cent at some seasons). These figures are probably maxima since they assume that no technological and societal adaptation will take place in the light of the changed climate. In any event, the changes of a few per cent in demand do not seem large, spread, as they will be, over tens of years; and should be easily handled by the electricity supply industries.

### What action can be taken to delay or reduce carbon dioxide-induced climatic change?

As discussed earlier, cutting the forests leads to a reduction in the amount of carbon stored in the terrestrial biosphere and a concomitant increase in the amount stored in other environmental reservoirs, including the atmosphere. Thus, one possible ameliorative action that man might consider taking to back-off increases in atmospheric $CO_2$ due to fossil fuel combustion is to plant more forests. However, this could only be a very partial answer to the problem since it turns out that vast land areas would need to be afforested. It has been calculated that in order to take up the approximately 2.5 GtC per year presently accumulating in the atmosphere from fossil fuel burning, an area equivalent to 10–20 per cent of the world's currently forested area would need to be planted. It seems extremely unlikely that such an enormous effort could be mounted, even assuming sufficient land suitable for forestation was available. Furthermore, even were the resources in terms of human effort and suitable land available, there would be very considerable pressure for them to be deployed to produce more food rather than for growing trees.

Even without planting more trees some additional uptake of atmospheric carbon dioxide may occur. As discussed earlier, crops kept under glass can be made to grow better with artificially high $CO_2$ levels, but the applicability of this result to the real environment, where factors other than carbon dioxide are likely to be limiting, is uncertain. One interesting suggestion is that with higher air levels of $CO_2$ some plants use soil moisture more efficiently. Although no quantitative global estimate of the significance of this effect seems to have been made, it does imply that land now marginal for plant growth may become more productive in an atmosphere containing higher amounts of carbon dioxide.

A technological solution to combat $CO_2$ emissions is to remove the gas from power station chimney stacks, in an analogous, but technically different, way to that employed to remove sulphur dioxide (discussed later). The problem is not just how to remove the $CO_2$ from the flue gas, but it also involves devising schemes for disposing of the removed carbon dioxide so that it does not return to the atmosphere. For example, it is pointless to scrub $CO_2$ from the stack gases into a solvent, such as water, if the solvent is subsequently discharged into a surface water environment. All that will happen is that the carbon dioxide removed will supersaturate the receiving water with respect to atmospheric concentrations and the dissolved $CO_2$ will merely degas back to the air. To overcome this problem the removed carbon dioxide must be deposited in a reservoir out of contact with the atmosphere on the longest possible time-scale. The deep oceans would seem to offer the best prospect from this point of view (although even here the waste will ultimately return to the surface on a time-scale of the order of a thousand years), but disposal by this means would clearly add very substantially to the costs of the overall operation. Even if sea-water itself is used as the scrubbing fluid, the costs both in money and energy terms are still very high. Some calculations indicate that the energy penalty would be such that more energy would be required to scrub the $CO_2$ than the power station itself would be producing! Somewhat less unfavourable is to use monothanolamine ($HO(CH_2)_2NH_2$) as the scrubbing fluid, with regeneration and reuse of the monoethanolamine and compression of the extracted $CO_2$ for transport to the coast and dumping in the deep sea. Removal and disposal of carbon dioxide costs money and also uses a significant fraction of the electricity output of the power station to run the extraction and pumping operations. Taken together the energy and financial penalties combine to increase the cost per unit of electricity generated by a factor of two to three. (These figures should be taken as rough estimates only, since the calculations contain many uncertainties).

However, even if the option was shown to be technically feasible (including the ultimate disposal of the removed $CO_2$) and the very substantial increase in the cost of electricity produced was acceptable, this solution would deal only with the approximate 30 per cent of the $CO_2$ from fossil fuel coming from power plants. The other 70 per cent comes from a myriad of generally small and often mobile sources from which it would be extremely difficult, if not impossible, to remove carbon dioxide effectively.

A more practical approach to reducing $CO_2$ emissions is to switch from carbon-based energy sources to those which produce less or no $CO_2$. All the renewable energy sources (wind, waves, solar, hydro) are essentially $CO_2$-free, as is nuclear power. Also, some fossil fuels produce less $CO_2$, per unit of energy output, as shown above in Table 2.5. Thus, a shift from coal towards a greater use of oil and gas would help to decrease $CO_2$ emissions; but would be limited by the relatively small proven reserves of oil and gas compared to coal. In passing, it is interesting to note that the reason why natural gas produces significantly less carbon dioxide than coal is because 60 per cent of the energy derived from gas comes from the conversion of hydrogen atoms in the paraffin molecules to $H_2O$, whereas in coal combustion the majority (80 per cent) of the energy comes from the oxidation of carbon atoms in the fuel to $CO_2$.

The last and probably most practical option of all in any attempt to reduce $CO_2$ input into the atmosphere is more efficient conversion and use of fossil fuels. If less wastage occurs in both the production and use of energy from these (and other) fuels then mankind can have the benefits the energy provides without the possible climatic disbenefits that rapidly rising levels of atmospheric carbon dioxide may provoke.

## Oxides of sulphur and nitrogen

It is clear from Table 2.4 (above) that after carbon dioxide, quantitatively the next most important emissions from power stations are the oxides of sulphur and nitrogen. Sulphur is emitted in the form of sulphur dioxide ($SO_2$) and nitrogen mainly as nitric oxide (NO). As we shall see, $SO_2$ can be subsequently oxidized in the atmosphere but this takes some time and occurs at considerable distances from the point of emission. In contrast, NO is rather readily and rapidly oxidized to nitrogen dioxide ($NO_2$) close to the release point. Because of this rapid conversion these two nitrogen oxides are often dealt with together and their sum (i.e. NO + $NO_2$) is then referred to as $NO_x$. This is the approach which will be taken here.

Power stations are not, of course, the only sources of man-made $SO_2$ and $NO_x$. Any combustion process using a fuel containing sulphur will result in the emission of $SO_2$, and similarly $NO_x$ will be formed either from oxidation of nitrogen contained in the fuel or, in high temperature burning, by combination of nitrogen and oxygen in the air used to support the combustion. It has been calculated, in fact, that in the UK, coal-fired power plants produce about two thirds of total emissions of sulphur dioxide. The situation for nitrogen oxides is somewhat different, with power stations contributing less than a half of total man-made emissions, the majority of the rest coming from motor vehicle exhausts. The relative importance of power stations for emissions of $SO_2$, compared with $NO_2$, is part of the reason why here we will give more attention to the former compared with the latter. Before considering in some detail the fate and effects of $SO_2$ (and $NO_2$) emissions on the environment, we will first look at the role of sulphur dioxide from fuel burning in the global cycle of the element sulphur. The equivalent global cycle for nitrogen will not be covered here. This is because there are significantly less data available and consequently less known about the nitrogen compared with the sulphur cycle.

In Fig. 2.18 one attempt at estimating the preindustrial and current sulphur flows in the environment is illustrated. Several other similar diagrams have been constructed and, although there are quantitative differences between them, the features of interest here are apparent in all of them. In these budget exercises the system at any one time is assumed to be at steady-state, with the sum of the flows into any reservoir (for example, the atmosphere) equalling the sum of the fluxes out.

In all the budgets, in order to achieve a balance it is necessary to have a substantial flux of volatile sulphur in some form coming from the oceans into the atmosphere. In early budgets this flux was obtained by difference, that is to balance the budget, and hydrogen sulphide ($H_2S$) was assumed to be the form in which the volatile sulphur was transferred. We now know that the main gaseous form of sulphur coming from the oceans is as the biologically produced gas dimethyl sulphide (DMS). This gas is made by certain types of phytoplankton living in ocean surface waters. Also the magnitude of the air–sea flux is now estimated by measurements of DMS made at sea, rather than just assumed to balance the budget.

Apart from being an important component of the global sulphur budget, the flux of dimethyl suphide out of the oceans also fulfils other important roles. For example, once in the atmosphere DMS is subject to chemical reaction leading to the formation of $SO_2$, among other things. This sulphur dioxide, of course, acts in exactly the same way as $SO_2$ emitted from the combustion of fossil fuels. Indeed, in parts of the world remote from anthropogenic influence, $SO_2$ and its further oxidation product sulphate, formed from the oxidation of DMS, are the chief sources of acidity found in rain and aerosols. Furthermore, even in parts of Europe close to the North Sea, which is a powerful emitter of DMS in spring and summer (Turner *et al.* 1988), the marine source of DMS is not insignificant as a provider of sulphur to the atmosphere, although man-made emissions from

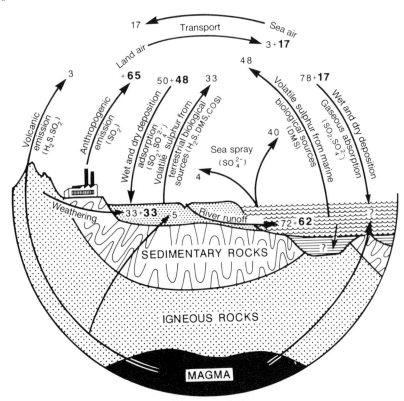

**Fig. 2.18** The global sulphur cycle. The inter-reservoir transfers are in MtS per year. Small type denotes transfers as estimated to have prevailed before man had a significant influence on the sulphur cycle. The bold figures give estimates of what man has added by various activities. Adapted from Granat *et al.* (1976) and Zehnder and Zinder (1980).

land are still the major source, particularly in winter when planktonic production of DMS is at a very low level.

Another possible role for dimethyl sulphide emitted from the oceans is as a climate regulation mechanism, as mentioned in an earlier section. The idea is that in the atmosphere over the remote open oceans, the chief source of sulphate-rich aerosols is by oxidation of marine-derived dimethyl sulphide. These fine sulphate particles serve as the main source of cloud condensation nuclei (CCN), that is the seeds on to which water vapour condenses to form cloud droplets. The fraction of energy reflected by clouds, called their albedo, is related to the number of condensation nuclei in unit volume of the cloud. Thus, with more CCN in the clouds a greater amount of incoming solar radiation is reflected away from the earth, causing it to cool. One of the key links in the theory is that if, for example, the surface earth warms due to man-made or other increase in atmospheric $CO_2$, then the oceanic plankton will produce more DMS which will lead to a greater number of sulphate CCN, and hence to a cooling trend. The overall result is that the $CO_2$-induced warming is being backed off by the increased cloud albedo. This idea has still to be proven since several links in the chain briefly described here have yet to be demonstrated to be of the right magnitude and to work in the correct sense. However, if proven, it is a very powerful theory since it describes a climate control system having negative feedback through the interaction of two of the most important natural cycles, those of carbon and sulphur.

As can be seen from Fig. 2.18, globally the natural flux of dimethyl sulphide out of the oceans is 4.8 MtS (megatons, expressed as sulphur; $1Mt = 10^6$ tonnes $= 10^{12}$ grams) each year. This is nearly as large as the anthropogenic input of sulphur to the atmosphere from burning fossil fuels (65 MtS per year). Another interesting point to note from the figure is that whereas prior to industrialization there was a net flux of sulphur from the sea to the land via the atmosphere ($17 - 3 = 14$ Mts per year), now this flux has been reversed and the land is a net contributor to the marine environment of 3 MtS per year.

The above discussion clearly indicates the profound influence man is having on the global sulphur cycle. However, anthropogenic emissions of sulphur dioxide are by no means evenly spread over the globe, being heavily concentrated in industrialized areas, such as North America, Europe and parts of Asia and South America. As an example, the yearly emission rates for the countries of Europe in the early 1980s are given in Table 2.6. This shows that the total emissions from Europe at almost 21 MtS per year constitute close to a third of total anthropogenic emissions of 65 Mt of sulphur per year (see Fig. 2.18), although the area of Europe represented by the countries listed in the table is only about 4 per cent of the land surface of the globe. To put the European emission figures in a rather more graphic way, they correspond to an input to the atmosphere of approximately 50 kilograms of sulphur (as $SO_2$) per year for each person living in the region. What is plain from the above discussion is that Europe is a very concentrated source of anthropogenic sulphur to the atmosphere and so an ideal area to examine the air chemistry and environmental effects of such emissions.

**Transport, air chemistry, and removal processes for emitted sulphur and nitrogen oxides**

For reasons of air quality, power station emissions to the atmosphere are invariably through a tall stack. The obvious purpose of such high chimneys is to ensure that the exhaust gas is well mixed, and so diluted, before it impinges on the ground downwind of the power station. For example, with

**Table 2.6**   Sulphur emission in Europe in the early 1980s in units of millions of tonnes of S (Mts) per year

| | | | |
|---|---|---|---|
| Albania | 0.050 | Ireland | 0.087 |
| Austria* | 0.215 | Italy* | 2.200 |
| Belgium* | 0.404 | Luxembourg* | 0.024 |
| Bulgaria* | 0.500 | Netherlands* | 0.240 |
| Czechoslovakia* | 1.500 | Norway* | 0.075 |
| Denmark* | 0.228 | Poland | 2.150 |
| Finland* | 0.270 | Portugal | 0.084 |
| France* | 1.800 | Romania | 0.100 |
| East Germany* | 2.000 | Spain | 1.000 |
| West Germany* | 1.815 | Sweden* | 0.275 |
| Greece | 0.352 | Switzerland* | 0.058 |
| Hungary | 0.750 | Turkey | 0.500 |
| Iceland | 0.006 | U.K. | 2.560 |
| | | Yugoslavia | 1.475 |
| | | TOTAL | 20.718 |

*Denotes a member of the Thirty Per Cent Club (countries which have agreed to reduce their sulphur emissions by 30% over the period 1980–1993).
*Source: Swedish Ministry of Agriculture (1982).*

a stack of height 220 metres the lower edge of the exhaust gas plume will not impinge with the ground until it is some 10 to 20 kilometres downwind of the point of emission, by which time dilution with ambient air will have reduced flue gas concentrations by a factor of about ten thousand.

Although the use of tall chimneys ensures that air quality close to the power station is largely unaffected by the presence of the plant, it does mean that we have to concern ourselves with what happens to the plume and the pollutants it contains at considerable distances from the point of release (as is also the case for low-level emissions, Fisher 1986). Fig. 2.19 illustrates the main processes of transport, removal, and chemical transformation of power station emissions after release.

In the atmosphere the major physical process affecting the plume is dispersion brought about by the prevailing wind. This mixing with ambient air occurs both horizontally and vertically. Dispersion in the vertical is largely controlled by the stability of the air with height. In general this means that upward mixing is confined to a layer seldom deeper than one kilometre. Air within this layer (often called the planetary boundary layer) is generally well mixed due to turbulent air motions. Above the well-mixed layer the air is stable with height, which inhibits vertical mixing and acts as a 'cap' on the polluted air below. In contrast, horizontal dispersion of the

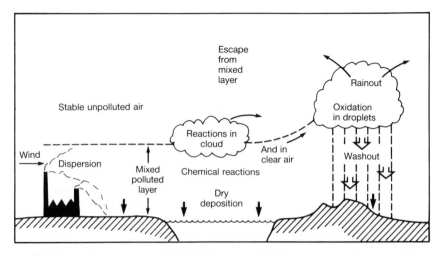

**Fig. 2.19** Processes of transport, removal, and chemical transformation of power station emissions after release (from Crane and Cocks 1987).

plume within the planetary boundary layer occurs over downwind distances of several hundreds of kilometres.

Although dispersion leads to dilution of the plume gases it does not decrease the total amount in the air. This can occur only through either chemical transformation, to which we will return later, or by removal from the atmosphere. The simplest form of removal is where the emitted gas molecules, e.g. $SO_2$ and $NO_x$, are taken up by land and water surfaces or by active uptake by vegetation. Such molecule by molecule removal of gases, without the intervention of atmospheric water droplets, is termed 'dry' deposition. In dry air this is essentially the only removal process possible. The rate at which dry deposition removes material from the plume depends on the concentration of the gas concerned and the speed at which it is taken up by the absorbing surface (quantified through a 'deposition velocity'). This latter varies between surfaces and also with the speed of the wind near the ground and the vertical stability of the atmospheric boundary layer. It has been calculated that for sulphur dioxide dry deposition alone will reduce concentrations in the mixed layer at a rate of a few per cent per hour. This means that in dry weather and under typical conditions a plume will still contain about half its load of sulphur dioxide one day after release, by which time it will be six hundred kilometres downwind of the source, assuming a wind speed of 7 metres per second. Since the rate of dry deposition of nitrogen oxides is generally slower than that for sulphur dioxide, removal by this mode will be even less effective for $NO_x$ than it is for $SO_2$. These examples clearly show the potential for long-range transport of power station gaseous emissions to distances of several hundreds of kilometres, at least under dry conditions.

However, studies of plumes have shown that removal rates significantly greater than those discussed above are often observed. One possibility is that chemical transformations of the plume constituents occur which may lead to their more rapid removal from the air mass. An example is the conversion of $NO_x$ to nitric acid vapour which rapidly deposits to land and water. However, although gas phase reactions do occur, they are generally either not very effective or do not lead to any enhancement in the rate of dry deposition.

A more potent set of transformations occur when the air mass receiving the power plant emissions is moist and has clouds containing water droplets. Then gases like sulphur dioxide and oxides of nitrogen are absorbed by the drops. This can lead to much more rapid removal to the ground of the chemicals contained in the air mass, i.e. 'wet' deposition of rain from clouds is a fast process compared to dry deposition of individual gas molecules. Wet deposition is composed of two components, called rainout and washout. The former is removal of pollutants from the cloud by falling water droplets, whereas washout refers to the scavenging of additional material by the drops as they fall between the base of the cloud and the ground. However, the capacity of cloud droplets to absorb gases is limited by the amount of gas which the drops can absorb, as well as by the rather small amount of water contained in the cloud drops (a typical cloud contains only about 1 gram per cubic metre of water in droplet form). The problem of limited solubility is overcome to a large degree because in solution chemical transformations of the sulphur and nitrogen oxides occur rather readily. For example, in aqueous solution oxidation of $SO_2$ to sulphate occurs much more easily than in dry air. Several oxidizing agents help to bring about the conversion. Atmospheric oxygen dissolved in the drops can oxidize $SO_2$ but the rate of the process is very slow unless trace metal catalysts of the reaction (e.g. iron and manganese) are also present. More potent oxidizing agents are ozone and hydrogen peroxide in the cloud droplets. The oxidation of sulphur dioxide to sulphate makes the cloud water considerably more acidic than it would otherwise have been. However, other reactions also occur in the drops and they can lead to some reduction in the acidity. For example, ammonia gas, a natural constituent of air but whose concentrations have probably been increased by agricultural activities, is very soluble in water and is thus readily taken up by cloud droplets where, because of its alkaline nature, it tends to neutralize some of the acidity. The final pH (a measure of acidity in which lower values correspond to greater acidity) of the rain is determined by the interaction of a large number of reactions of the types briefly described here. Despite some neutralization by ammonia, it is clear that rain from clouds downwind of power plants and large industrial and urban areas has a lower pH (i.e. more acid) than that in pristine parts of the globe. In broad terms rain unaffected by man-made emissions has a pH in the region of 5 (for example Galloway and Gaudry's (1984) data from Amsterdam Island in the middle of the Indian Ocean), whereas rain in Europe and other highly industrialized regions regularly has pH values in the region of 4.5 or lower. Since pH is a logarithmic scale, this pH difference corresponds to at least a three-fold change in acidity between clean and polluted areas. It should be noted that even in unpolluted regions the rain is still acidic (i.e. has a pH of less than 7—the neutral point on the pH scale). As mentioned earlier, this 'natural' acidity largely derives from $SO_2$ formed from the oxidation of dimethyl sulphide produced in sea-water by marine phytoplankton.

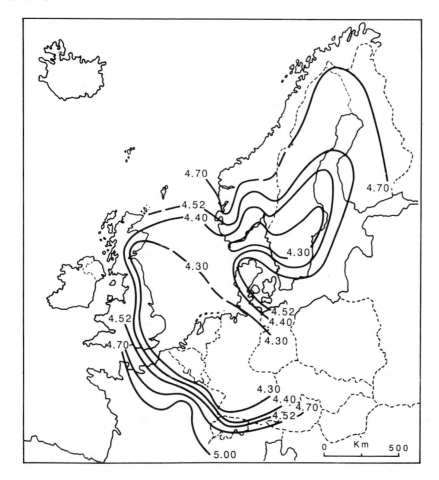

**Fig. 2.20** Variations in acidity of rainfall over Europe, 1974 (from Park 1987, based on OECD data).

Figure 2.20 shows the distribution of rain-water pH over north-western Europe and gives some idea of its spatial variability. Values are generally higher in the industrial heartlands of Germany, the Low Countries and eastern UK and lower in the less populated/industrialized areas to the north and south. Data for temporal trends in rain-water acidity are harder to find since reliable measurements which may be compared over a period of years are scarce. One such attempt is shown in Fig. 2.21 which is for rain-water pH in the eastern USA in the mid 1950s compared with the early 1970s. A clear shift to lower (more acidic) pH values is apparent in this period.

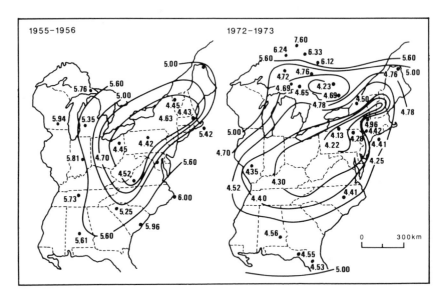

**Fig. 2.21**  Increase in acidity of rainfall over the Eastern United States between 1955–6 and 1972–3 (from Park 1987).

Several mathematical models have been constructed to simulate the deposition (both wet and dry) of sulphur over Europe from industrial sources. The output from one of these models is shown in Fig. 2.22. Maximum deposition occurs in the broad industrialized band stretching from England across Germany and into Eastern Europe. The general agreement between the contour distribution in the model predictions and the distribution of rainfall pH (Fig. 2.20) is reassuring but must be treated with caution. For example it is known that the model underestimates the amount of sulphur deposited in rain in the wetter regions of Europe. Also, model predictions do not quantitatively balance with measured rates of deposition. This mis-match is usually attributed to 'background' sources of sulphur not included in the models which invariably deal only with

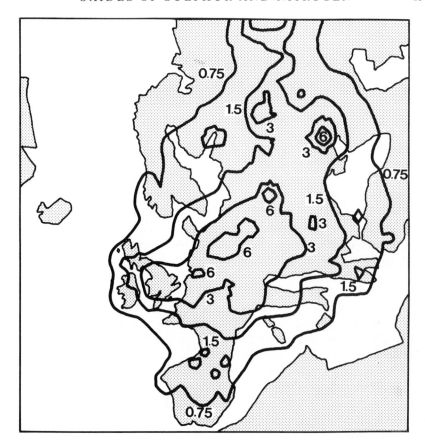

**Fig. 2.22** Estimated sulphur deposition in g m$^{-2}$ yr$^{-1}$ for 1980 over Europe (from Crane and Cocks 1987).

industrial emissions. The origin of this background is unclear but it could be due *inter alia* to very long-range transport (e.g. from North America) or biogenic sulphur emissions from sea-water. Some support for this latter suggestion comes from the fact that the greatest discrepancy between the observed depositions and the model predictions occurs in the spring when marine phytoplankton activity is at its maximum. Improved modelling of this type is an important goal since the models serve many important purposes. For example, they are almost the only way of attributing quantitatively the acidity deposited in one area to its various places (countries) of origin, clearly a highly political issue. Further, models are vital in predicting the effects in terms of changes in deposition of acidity which will arise as a result of reducing industrial emissions of sulphur and nitrogen oxides. The effects are not necessarily linear (i.e. halving emissions may not lead to a fifty per cent decrease in deposition), and sophisticated models are required to predict them.

### Environmental effects of sulphur and nitrogen oxides and acid rain

It is clear from the discussion in the previous section that in industrialized/ urbanized regions the acidity of rain can be significantly increased as a result of the combustion of fossil fuels. Here we will be concerned with what effects such increased acidity can have on various parts of the natural and built environment. Natural reservoirs to be considered include soils, rivers and lakes, crops and forests. Most work on the impact of acid rain on buildings has been carried out on cathedrals and other such prominent ancient structures.

When rain falls on to land the most common surface on which it impinges is soil and its covering of vegetation. The soil is well used to acid inputs since even clean rain is acidic and microbial breakdown of organic plant material produces quite naturally further acidity *in situ*. Thus, rain carrying additional pollution-derived acid does not create a new situation, it just amplifies an existing one. What happens to any additional acid, and its effects, depend very much on the types of minerals making up the soil, as well as their amount, i.e. the thickness of the soil cover. Where the soil is rich in minerals composed of, for example, calcium carbonate (chalk and limestone), which react readily with acid, the incoming acidity is easily neutralized. The water flowing off and through such soils is generally alkaline and the additional acidity does not have a deleterious effect on the soil itself. Alumino-silicate minerals, such as feldspars, can similarly react with and neutralize acidity but their ability in this respect is substantially less than that of carbonate minerals. A third type of solid component common in soils is clay minerals whose important property in the present context is the ability to exchange ions between themselves and the surrounding soil water. Thus, an increase in the acidity (concentration of hydrogen ions, $H^+$) of the soil water leads to uptake of $H^+$ ions and release of an equivalent amount of positively charged ions (cations) previously held by the clay minerals. Then the water draining from the soil is enriched in cations such as calcium ($Ca^{2+}$), aluminium ($Al^{3+}$) and also trace metals. Organic matter in soils acts in a similar way to clay minerals in terms of ion exchange ability. Where the soil cover is thin or where it lacks the types of minerals and organic matter discussed above, it will have a much reduced ability to withstand extra inputs of acid, and the composition of the rain-water as it percolates through such soils will be relatively little changed. This is the situation in parts of Scandinavia and areas of the Canadian Shield region in North America.

There are several ways in which soil fertility is affected by increases in acidity from rain. Obviously the pH of the soil will tend to decrease, except in very well buffered horizons (i.e. those containing minerals capable of neutralizing added acid), and this will disadvantage some plant species, but may well favour others. Further, the release of soluble forms of aluminium and trace metals may prove toxic to some crops. Increased acidity will also affect soil, fauna, and flora. Studies indicate that in acidified soils rates of decomposition of organic matter by bacteria and other micro-organisms are slowed down, and there are fewer earthworms but more fungi. However, the effects of acid rain on soils are not always detrimental. For example, in soils which are poor in nitrogen and, less commonly, sulphur the enhanced inputs of these important plant nutrients via rainfall deposition of anthropogenically produced nitrogen and sulphur oxides may be of positive benefit.

The clearest impacts of acidified rain on the environment are probably to be found in rivers and lakes. In regions with well developed soils which can neutralize acidity any impact of acid inputs on freshwaters derived from water which has drained through such soils is likely to be small. As might be predicted, the greatest impacts occur where the soil cover is thin and relatively lacking in minerals capable of reacting with acids. In these cases significant decreases in water pH are to be expected, and are indeed found, in regions downwind of industrial/urban complexes.

Probably the best evidence for decreases in freshwater pH with time since industrialization is that presented by Batterbee et al. (1985) from their studies of the species of diatoms found in the sediments of several lakes. Diatoms are phytoplankton which build their structures largely of silica which they scavenge from the water in which they live. It is known that individual species of diatoms are very specific as to the acidity of the water in which they can grow. Thus, if the pH of the lake or river changes for whatever reason, the types of diatoms that can prosper in that water will reflect the alteration in acidity status. Figure 2.23 shows some results for Loch Enoch in south-west Scotland. From the assemblages of diatom species at different depths in a sediment core from the lake, the pH of the lake-water in which the diatoms lived is inferred. By dating the core using the radio-isotopes it contains, a chronology of water acidity is obtained. This lake appears to have had a pH that was reasonably constant, at 5.0–5.5, up until about 1830. Later than this, with increase in industrial and urban development, the incidence of acid-loving diatoms has increased progressively over the years and the water pH inferred has dropped to below 4.5 in recent times.

Such changes in freshwater pH of half to one unit have been documented for several lakes in areas downwind of and susceptible to acidified inputs. Drops in pH of this magnitude lead to significant effects on other organisms living in the receiving waters. Figure 2.24 gives some indication of the sort of changes that occur. Fish, and in particular their juvenile forms, are especially susceptible to increases in acidity. Below pH 5.5 most fish populations show signs of decline and few survive below 5.0. An

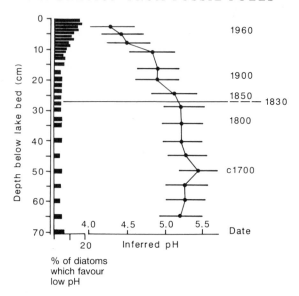

**Fig. 2.23** Water pH of Loch Enoch (SW Scotland) as a function of time, inferred from diatom species at various depths in the bottom sediments (after Batterbee *et al* 1985).

analysis of the status of almost 1700 lakes in southern Norway with respect to their fish populations illustrates the point. In the study region the lakes were divided into the following three categories according to the populations of fish they supported: good fish population, sparse fish population, and no fish. Figure 2.25 shows for each of the three categories the percentage of lakes having particular water pHs. It is clear that high pH corresponds to good fish populations, but for water pH below 5.0 a large percentage of the lakes have either sparse fish populations or none at all.

It is worth briefly noting two things at this point. One is that it is not necessarily the acidity ($H^+$ ion) *per se* which brings about the deleterious effects on fish. There is now considerable evidence that aluminium in low pH water is in a form that can cause harm to fish. As discussed earlier in this section, aluminium gets into the water as a result of various chemical and physical processes as water percolates and reacts with soil particles. Secondly, the temporal changes in lake-water acidity described above are not necessarily caused only by increased rainfall acidity. In some areas, at least, changes in land use, in particular forest cutting and alterations in agricultural practice, may also have played a contributory role.

Another possible impact of acidified rain is on crops and plants. The effect of increased acidity on soils has already been discussed; here we are more concerned with whether sulphur and nitrogen oxides and low pH rain can have a direct adverse effect on plant growth. Laboratory experiments with crops exposed to elevated $SO_2$ levels do indeed show that damage can occur, but only at concentrations significantly higher than ambient. Indeed, at low added levels of sulphur and nitrogen oxides there is some evidence for enhancement in plant growth, due to the extra nutrient provided by the 'pollutant' gases. In the past, before the advent of the tall chimney policy for power stations, sulphur and nitrogen oxide gas levels may well have been high enough for crop damage to have occurred. Thus, tall stacks have brought one problem under control but have at the same time exacerbated the potential for environmental effects 'at a distance', as discussed earlier.

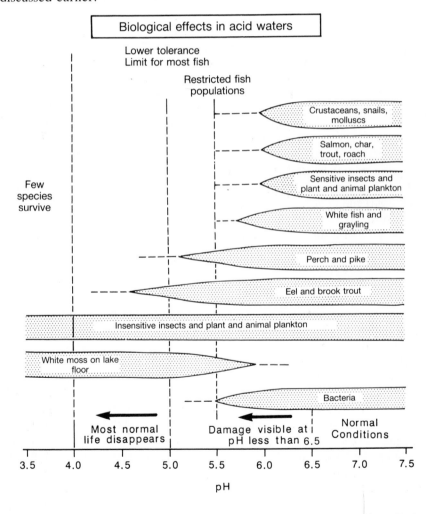

**Fig. 2.24** Changing tolerance of freshwater species as water pH falls (from Park 1987).

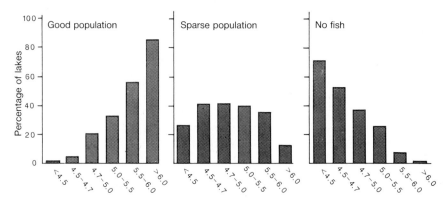

**Fig. 2.25** Fish populations in a survey of 1679 lakes in southern Norway as a function of water pH (from Park 1987 after Leivestad *et al.* 1976).

Much has been heard in recent years on the so-called 'forest dieback' problem. Trees, particularly in Germany, show signs of damage and death in large numbers. Initially this problem was attributed to acid rain and according to some the elevated levels of soluble aluminium in soils resulting from increased soil acidity. Although acidification may play some role, it is now thought that other atmospheric pollutants (such as ozone and ammonia) and deficiencies of certain elements in the soil (e.g. magnesium) are also likely to be important (Blank *et al.* 1988).

The acidic products of fossil fuel combustion can also have detrimental effects on building-stone. This leads to the black, grimy appearance of many buildings in industrial cities due to soot and sulphur dioxide from previous fuel burning. Up until relatively recently coal was burnt in a multitude of domestic and industrial fires and furnaces having low-level chimneys. Now much more energy is supplied from power stations with their tall stacks or is produced locally but from fuels producing little smoke and less $SO_2$.

However, the result of burning fossil fuels is not only to blacken buildings, which effect can be reversed by cleaning. More fundamentally, $SO_2$ and acidified rain will lead to enhanced corrosion of the building-stone itself. If the building is made of limestone or marble then these minerals will react with sulphuric acid formed from deposited $SO_2$ to form calcium sulphate. This compound is much more friable than the original calcium carbonate material and so tends to peel off the face of the stone, thus exposing the underlying material to further corrosion. Where the building is made of sandstone then the attack is on the calcium carbonate or iron oxide which holds the individual quartz grains together. Although it might be thought that corrosion of buildings should be on the wane in view of the sort of decreasing trends in $SO_2$ pollution in urban areas discussed above, it is not well established that this is the case. What observations there are do not strongly support a decrease in stonework corrosion in recent times. It may be that although $SO_2$ levels have decreased, other gases such as ozone and oxides of nitrogen (both arising from motor vehicle emissions) have increased as the number of cars in urban centres has grown, and that these are now the compounds that corrode the fabric of buildings.

## Possible health effects of power station emissions

It is well established that burning coal in open grates can have deleterious impacts on human health. The more obvious short-term effects are on the respiratory system and are produced by smoke and in particular, sulphur dioxide and sulphuric acid derived from its oxidation. London 'smogs' and their resultant excess deaths are infamous. The best-known example is in December 1952 when during one week 4000 deaths occurred over the expected number, largely due to respiratory failure produced by high levels of smoke and $SO_2$. Such events led to the Clean Air Act of 1957 and a very substantial improvement in air quality in London and other UK cities.

With the phasing out of such an obvious source of pollution as the open domestic fire, attention has turned to the possible health hazards arising from power station emissions. As we have seen, considerable effort is made to remove particulates from power station exhausts, but they still emit large quantities of sulphur and nitrogen oxides. Because the emission is from a tall stack, the large dilution this affords keeps ground-level concentrations low. As with exposure to most pollutants at low concentrations, estimation of the risk to health is extremely difficult. Fremlin (1987) quotes a range of between 0.1 and 77 deaths per year for each 1000 megawatt of power station capacity; the large range emphasizing the uncertainties in the calculation. Within this range he settles for a most likely value of three deaths per year per 1000 MW, giving a total of 120 per year for the whole UK.

Another potential health hazard of power generation is cancer induced by substances such as polycyclic aromatic hydrocarbons. These are produced by incomplete burning of fossil fuels and are known carcinogens, particularly for lung cancer. There are many sources of these compounds, since they are invariably formed when fossil fuels are burnt. It has been estimated that in the UK of the order of 1000 deaths per year may be attributed to this cause. Given their widespread occurrence, it is very difficult to attribute these deaths to particular emitters, such as cars, local sources, and power plants. Fremlin's best estimate for power stations is one death per year per 1000 MW of coal and oil fired plant, over and above the three deaths mentioned earlier from oxide emissions.

## Remedial action to combat the environmental effects of sulphur and nitrogen oxides

As discussed in the previous section, oxides of sulphur and nitrogen emitted from fossil fuel-fired power stations can have a variety of detrimental environmental impacts. In view of these, considerable thought and effort has gone into investigating how they might, preferably, be avoided or at least ameliorated. The various strategies are discussed in the following paragraphs.

The most obvious possibility is to burn less fossil fuel. This can be done either by the more efficient conversion and use of coal and oil and/or by switching to power sources which do not involve emission of $SO_2$ and $NO_x$. The latter option essentially means a greater reliance on nuclear power and the renewable sources of energy, such as wind, waves, solar, and hydro. This topic has already been raised in the context of reducing emissions of carbon dioxide from power stations and will not be discussed further here.

Another way to proceed is not to try to reduce emissions of $SO_2$ and $NO_x$ but to try to counteract their harmful environmental consequences. This is not a generally favoured route since it is extremely difficult, if not impossible, to cancel out all the adverse impacts. However, it may have some utility as a short-term palliative while other more radical solutions are sought and implemented. For example, lime (calcium carbonate) has been added to many lakes in Sweden to neutralize acididty arising from low pH rain and runoff. Some success has been achieved in this way in restoring fish populations.

More fundamental solutions naturally involve tackling the problem at its source, i.e. at the power station. Before these are discussed it is worth making two points.

The first is that power plants are easily identified, large sources of a variety of atmospheric pollutants. As such it is conceptually easy to see why reducing their emissions is a first priority in any clean-up operation. However, in the UK about a third of the $SO_2$ and more than half of the $NO_x$ emissions come from sources other than power stations. As most of these other sources are small and often mobile (cars, lorries, etc.) controlling their emissions is a much more difficult task.

A second point worth noting is the probable lack of linearity in any $SO_2$/ $NO_x$ abatement control policy. What this means is that even if power station emissions of, for example, $SO_2$ were to be reduced by, say, 50 per cent, this would not necessarily lead to a reduction of the same percentage in deposited sulphur at any particular locality downwind. The non-linearity arises in several ways. Power plants are not the only sources, and other man-made and natural inputs will continue to emit at the previous rate. In addition the rate at which nitrogen and sulphur oxides are chemically transformed to sulphuric and nitric acid in the atmosphere is very much dependent on the presence of other substances, e.g. ozone in the oxidation of $SO_2$ to sulphate in rain. Since the concentration of many of these other species has essentially nothing to do with what happens inside the power plant, a linear relationship between reduction in power station emissions and deposition in specific areas is hardly to be expected.

Bearing these caveats in mind we now examine other means of reducing emissions of $SO_2$ and $NO_x$ from power stations. A fairly obvious way of reducing sulphur dioxide output is to burn fossil fuel containing less sulphur. In the UK the coal mined in different areas shows a considerable range of sulphur contents, as shown in Table 2.7. By using coal only from certain regions it would in principle be possible to reduce $SO_2$ output. However, this would increase transport costs since coal would have to be moved further between pit and power station. Further, in the longer term the pits producing low sulphur coal would become depleted more rapidly thus creating greater problems in the future. Another simple solution is to wash the coal more than is done already before it is burnt. This leads to

**Table 2.7**  Sulphur content of United Kingdom coal, 1982–3

| Source area | Average sulphur content (%) | Saleable output (million tonnes) |
|---|---|---|
| Scottish | 0.70 | 6.6 |
| South-west opencast | 0.90 | 2.1 |
| South Wales | 0.95 | 6.9 |
| Scottish opencast | 0.95 | 2.8 |
| South Notts | 1.37 | 8.3 |
| North Notts | 1.48 | 12.4 |
| Doncaster | 1.48 | 6.8 |
| North-east | 1.49 | 12.4 |
| Western | 1.52 | 10.8 |
| South Midlands | 1.56 | 8.2 |
| North-east opencast | 1.56 | 3.1 |
| South Yorks | 1.59 | 7.3 |
| North Derbyshire | 1.69 | 8.1 |
| Barnsley | 1.85 | 8.1 |
| North Yorks | 1.92 | 8.4 |
| Central east opencast | 1.96 | 3.2 |
| Central west opencast | 2.07 | 2.5 |
| North-west opencast | 2.34 | 1.0 |
| TOTAL | 1.51 | 119.0 |

*Source: Dudley et al. (1985), Table 6*

removal of some of the sulphur in the form of pyrite ($FeS_2$) but overall the washing procedure is not very efficient and leads to $SO_2$ emission reductions of only about 10 per cent. Furthermore, neither burning low sulphur fuel nor coal washing do anything to reduce emissions of nitrogen oxides.

A more technically advanced option which reduces both $SO_2$ and $NO_x$ emissions is called 'fluidized bed combustion'. In this technology a mixture of powdered coal and limestone is burnt in the power station boiler. A stream of air passes upwards through the bed of mixed coal and lime and at the optimum air-flow the bed resembles a bubbling liquid (the 'fluidized bed'). The advantages of this arrangement are threefold:

1. The combustion process is very efficient.

2. The limestone reacts with any sulphur dioxide formed so that only a relatively small amount of it goes up the chimney.

3. The combustion temperature is lower than in a conventional burner so that fewer nitrogen oxides are generated.

For the present, fluidized bed combustion is a relatively untried technology, particularly on large power plants; however, it may well prove to be a successful option for the future.

Another technical solution to the sulphur dioxide emission problem is to remove the $SO_2$ from the exhaust gas stream before it exits from the power plant stack. The most widely used technique for doing this is called 'flue gas desulphurization' (FGD). The commonest way FGD is used is in the so-called limestone/gypsum system. In this the flue gases are passed through a slurry of limestone in water and the $SO_2$ reacts with the lime to form gypsum ($CaSO_4$). The process has an efficiency for removing sulphur dioxide of between 70 and 90 per cent. It is a well-tried system and so should present few technical problems. Flue gas desulphurization is not a new concept, for example Battersea power station in London had FGD fitted and removed $SO_2$ with 80–90 per cent efficiency for most of its working life, although in this case the $SO_2$ absorber was water from the River Thames, not limestone.

There are, of course, problems with FGD. It does little or nothing to remove nitrogen oxides, and requires considerable amounts of limestone and produces large quantities of wet gypsum slurry. The latter can be dewatered and dried, after which it can either be sold to the building trade (mainly for making plasterboard and wall plaster) or has to be disposed of. Clearly sources of limestone have to be available close to the plant or it has to be brought in from further away, thus adding to the transport costs. The amounts of additional material which have to be transported to the power station and ultimately disposed of when FGD is fitted are not trivial by any means. In Fig. 2.26 the problem is illustrated for the Drax power station in

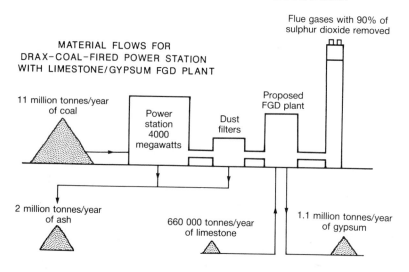

**Fig. 2.26** Material flows for Drax coal-fired power station with limestone/gypsum FGD plant (from CEGB 1988).

Yorkshire, to which the CEGB has recently announced its intention to fit FGD. Although this is admittedly a large plant (4000 MW), every year more than half a million tons of limestone are required and more than a million tons of gypsum are produced which has to be sold or disposed of. Once fully operational (planned for 1995) the FGD plant should remove more than 90 per cent of the $SO_2$ at a total capital cost of £400 million. Taken together the capital and additional running expenses will probably increase the cost of generating electricity at Drax by about 10 per cent.

To complete this section and to put $SO_2$ emissions and their removal into a historical and future context, it is useful to examine Fig. 2.27. This shows total UK emissions of $SO_2$ since 1860 as well as the amount of the total that is attributable to power stations. Total emissions have increased for most of the period, generally in line with the level of economic activity. However, there was a sharp downturn in the 1960s following implementation of the Clean Air Act of 1957, which regulated the burning of fossil fuels in urban areas. From the 1920s, when the electricity supply industry became established, until about 1970 the curve for $SO_2$ output by power stations closely follows the trend of total emissions. After that, the trend in $SO_2$ outputs from power plants is generally downwards but not as marked as for total emissions. The future projections indicated in Fig. 2.27 are for a continuing decline in $SO_2$ emissions from power stations as more FGD plant is installed on existing stations. A steeper decline is predicted for the late 1990s and early part of the next century as new power plants are

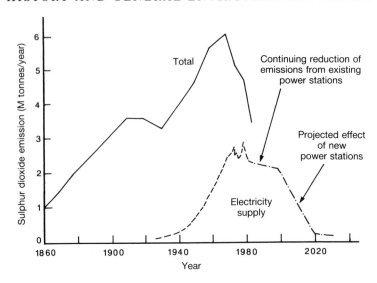

**Fig. 2.27**   UK SO$_2$ emissions as a function of time (from Chester 1987).

constructed which are either nuclear, and so free from sulphur emissions, or conventional stations with sulphur removal built in from the outset.

## References

Batterbee, R.W., Flower, R.J., Stevenson, A.C., and Rippey, B. (1985). What evidence is there for acid rain affecting freshwater ecosystems in the past? In *Report of the Acid Rain Inquiry*, pp. 89–95. Scottish Wildlife Trust, Edinburgh.

Blank, L.W., Roberts, T.M., and Skeffington, R.A. (1988). New perspectives on forest decline. *Nature* 336, pp. 27–30.

Bolin, B., Houghton, R.A., and Moore, B. (1985). *Changing forests and the CO$_2$ concentration in the atmosphere*. United Nations University.

Bolin, B. (1986). How much CO$_2$ will remain in the atmosphere? In *The greenhouse effect, climatic change, and ecosystems* (ed. B. Bolin, B.R. Doos, J. Jager, and R.A. Warrick) pp. 93–155. Wiley, Chichester.

Callendar, G.S. (1958). On the amount of carbon dioxide in the atmosphere. *Tellus* **10**, pp. 243–8.

CEGB (1980). *Submission to the Commission on Energy and the Environment-Topic 4 Combustion Residues*. CEGB, London.

CEGB (1987). *CEGB Annual Report 1986/87*.

CEGB (1988). *Drax power station, proposed flue gas desulphurization plant*. CEGB, London.

Charlson, R.J., Lovelock, J.E., Andreae, M.O., and Warren, S.G. (1987). Oceanic plankton, atmospheric sulphur, cloud albedo and climate. *Nature* **326**, pp. 655–1.

Chester, P.F. (1987). Acid rain — a prognosis. In *CEGB Research* No. 20, pp. 62–4. CEGB, London.

Clarke, A.J. (1981). *Disposal of ash from UK power stations: environmental problems and answers.* Paper presented to Chemical Institute of Canada, Halifax, Nova Scotia, June 1981.

Cohen, A.V. and Pritchard, D.K. (1980). *Comparative risks of electricity production systems: a critical survey of the literature.* Health and Safety Executive.

Crane, A.J. and Cocks, A.T. (1987). The transport, transformation and deposition of airborne emissions from power stations. In *CEGB Research* No. 20, pp. 3–15. CEGB, London.

Crane, A. and Liss, P. (1985). Carbon dioxide, climate and the sea. *New Scientist* No. 1483, pp. 50–4.

Dudley, N., Barrett, M. and Baldock, D. (1985). *The acid rain controversy.* Earth Resources Research, London.

Fisher, B.E.A. (1986). The effect of tall stacks on the long range transport of air pollutants. *Air Pollution Control Association Journal* **36**, pp. 399–401.

Fremlin, J.H. (1987). *Power production, what are the risks?* Clarendon Press, Oxford.

Galloway, J.N. and Gaudry, A. (1984). The composition of precipitation on Amsterdam Island, Indian Ocean. *Atmospheric Environment* **18**, pp. 2649–56.

Granat, L., Hallberg, R.O., and Rodhe, H. (1976). The global sulphur cycle. In *Nitrogen, phosphorus and sulphur — global cycles* (ed. B.H. Svensson and R. Soderlund) pp. 89–134. Swedish Natural Science Research Council, Stockholm.

Hansen, J., Johnson, D., Lacis, A., Lebedeff, S., Lee, P., Rind, D., and Russell, G. (1981). Climate impact of increasing atmospheric carbon dioxide. *Science* **213**, pp. 957–66.

Hansen, J., Fung, I., Lacis, A., Rind, D., Lebedeff, S., Ruedy, R., Russell, G., and Stone, P. (1988). Global climate changes as forecast by Goddard Institute for Space Studies three-dimensional model. *Journal of Geophysical Research* **99**, pp. 9341–64.

Hart, A.B. and Lawn, C.J. (1977). Combustion of coal and oil in power station boilers. In *CEGB Research* No. 5, pp. 4–17. CEGB, London.

HMSO (1981). *Coal and the environment.* Commission on Energy and the Environment.

HMSO (1983). *The Government's response to the Commission on Energy and the Environment's Report on Coal and the environment,* Cmnd 8877.

HMSO (1987). *The Coal Industry.* House of Commons Energy Committee.

Houghton, R.A. (1984). Estimating changes in the carbon content of terrestrial ecosystems from historical data. In *Global carbon cycle* Sixth ORNL Life Science Symposium.

Kellogg, W.W. (1978). Global influences of mankind on the climate. In *Climatic Change* (ed. J. Gribbin). Cambridge University Press.

Klein, D.H., Andren, A.W., Carter, J.A., Emery, J.F., Feldman, C., Fulkerson, W., Lyon, W.S., Ogle, J.C., Talmi, Y., Van Hook, R.I., and Bolton, N. (1975). Pathways of thirty-seven trace elements through coal-fired power plant. *Environmental Science and Technology* **9**, pp. 973–9.

Lagadec, P. (1982). *Major technological risk.* Pergamon, Oxford.

Leivestad, H., Hendrey, G., Muniz, I.P., and Snekvik, E. (1976). *Impact of acid precipitation on forest and freshwater ecosystems.* SNSF, Norway.

Liss, P.S. and Crane, A.J. (1983). *Man-made carbon dioxide and climatic change: A review of scientific problems.* Geo Books, Norwich.

Machta, L. (1979). Atmospheric measurements of carbon dioxide. In *Workshop on the global effects of carbon dioxide from fossil fuels* (ed. W.P. Elliott and L. Machta) pp. 44–50. U.S. Department of Energy.

Manabe, S. and Wetherald, R.T. (1980). On the distribution of climate change resulting from an increase in $CO_2$ content of the atmosphere. *Journal of Atmospheric Sciences* **37**, pp. 99–118.

Marland, G. and Rotty, R.M. (1980). Atmospheric carbon dioxide: What to do? *Consensus* **2–3**, pp. 41–52.

Natusch, D.F.S. (1978). *Potentially Carcinogenic Species Emitted to the Atmosphere by Fossil-fuelled Power Plants.* Environmental Health Perspectives **22**, 79.

Neftel, A., Moor, E., Oeschger, H., and Stauffer, B. (1985). Evidence from polar ice cores for the increase in atmospheric $CO_2$ in the past two centuries. *Nature* **315**, pp. 45–7.

OECD (1983). *Coal, environmental issues and remedies.*

OECD (1985). *Environmental effects of electricity generation.*

Park, C.C. (1987). *Acid rain: rhetoric and reality.* Methuen, London.

Rotty, R.M. (1977). Global carbon dioxide production from fossil fuels and cement, A.D. 1950–A.D. 2000. In The fate of fossil fuel $CO_2$ in the oceans (ed. N.R. Andersen and A. Malahoff) pp. 167–81. Plenum, New York.

Rotty, R.M. (1980). Uncertainties associated with global effects of atmospheric $CO_2$. *Science of the Total Environment* **15**, pp. 73–86.

Rotty, R.M. (1983). Distribution of and changes in industrial carbon dioxide production. *Journal of Geophysical Research* **88**, pp. 1301–8.

Spedding, D.J. (1974). *Air pollution.* Clarendon Press, Oxford.

Swedish Ministry of Agriculture (1982). Environment '82 Committee of Stockholm conference on Acidification of the Environment. *Acidification today and tomorrow*, p. 43, Stockholm.

Turner, S.M., Malin, G., Liss, P.S., Harbour, D.S., and Holligan, P.M. (1988). The seasonal variation of dimethyl sulphide and dimethylsulphoniopropionate concentrations in nearshore waters. *Limnology and Oceanography* **33**, pp. 364–75.

Whittacker, R.H. and Likens, G.E. (1975). The biosphere and man. In *Primary productivity of the biosphere* (ed. H. Lieth and R.H. Whittaker) pp. 305–28. Springer-Verlag, Berlin.

Zehnder, A.J.B. and Zinder, S.H. (1980). The sulphur cycle. In *The handbook of environmental chemistry* (ed. O. Hutzinger) Vol. 1 Part A, pp. 105–45. Springer-Verlag, Berlin.

# 3

# GENERATION FROM NUCLEAR FISSION

## Thermal and fast reactors

The fundamental process on which nuclear power depends—the fission of uranium—is the cause both of its environmental advantages and of its chief disadvantage (fusion is discussed in Chapter 4). The fission process involves the release of very large amounts of energy. The energy released when one atom of carbon (coal) combines with one molecule of oxygen is $7 \times 10^{-19}$ J. The energy released from the fission of one uranium nucleus is $3.2 \times 10^{-11}$ J, about fifty million times greater. In terms of weight of fuel, the total fissioning of one ton of uranium would therefore be equivalent to the burning of 2.7 million tons of coal. Because so little uranium can produce so much energy, the transport and storage of fuel presents few problems. Less land is needed than for fossil-fired stations because there is no need for a coal store or oil tanks, power stations do not have to be sited near to the source of fuel or by railway lines or ports, and the quantity of waste to be disposed of is very small compared, in particular, with that produced by coal-fired stations. However, the fission process also results in the production of radioactive materials and very considerable resources have to be devoted to the protection of workers and members of the public from the radiation that these materials emit.

The dramatic developments that led in less than fifty years from the discovery of radioactivity by Becquerel in 1896 to the first self-sustaining nuclear chain reaction in Chicago on 2 December 1942 are well known (Williams 1978). The transmutation of elements, the alchemists' dream, was first demonstrated in the laboratory by Rutherford in 1919. The neutron was discovered by Chadwick in 1932. In 1934 Fermi bombarded a range of elements with neutrons. He used uranium, the heaviest naturally occurring element, hoping to create new, even heavier elements. The results were complex and difficult to interpret. It was not until 1938 that the clue to what had happened emerged. The bombardment had indeed resulted in the creation of other elements, but they were lighter, not heavier than uranium. Curie and Savitch identified lanthanum and Hahn and Strassman identified barium among the products. Frisch and Meitner explained the results in terms of fission of the uranium nuclei. The key to harnessing these apparently esoteric reactions was the discovery by Joliot, von Halban and Kowarski in 1939 that the neutron-induced fission of uranium was accompanied by the emission of an average of three secondary neutrons; clearly a self-sustaining chain reaction was possible and uranium could be used to produce heat and, under the right conditions, an unprecedently powerful explosion. It was the latter possibility that led to the setting up of the Manhattan Project to develop the atomic bomb.

It turns out that of the two naturally occurring uranium isotopes only uranium-235 splits readily when bombarded by neutrons; uranium-238 is far more likely to absorb the neutrons. But natural uranium consists almost entirely of U-238, with only 0.7% of U-235; the nuclear properties of these two isotopes are such that natural uranium alone cannot sustain a chain reaction. In 1939 Bohr and Wheeler found that if the fast neutrons that are emitted when a uranium nucleus splits are slowed down they are more likely to cause further fissions. What was needed, therefore, was some way of slowing down (or 'moderating') the neutrons without absorbing them. The ideal moderator was heavy water($D_2O$) but this was only available in very small quantities at the time. The next best material was graphite, and it was a 'pile' of graphite and natural uranium that Fermi assembled under the stand of the Chicago University baseball stadium, under conditions of great war-time secrecy. The pile also contained rods of neutron-absorbing material—'control rods'. As these were gradually withdrawn the neutron intensity increased till the pile became 'critical': the chain reaction became self-sustaining. Fermi's simple pile confirmed the calculations that had suggested that nuclear power was possible but a practical reactor also needed cooling to remove the heat generated and a biological shield to protect the operators from the neutrons. The first such reactor was built at Oak Ridge, Tennessee in 1943. It was air-cooled and had a heat output of 1800 kW. It contained the essential features of all future thermal reactors (so-called because the neutrons are slowed down till their velocities are in thermal equilibrium with the molecular velocities of the surrounding materials) that is fuel, moderator, control rods, coolant, and shielding. The next stage was to build large-scale reactors for the production of plutonium (see p. 80). Three reactors were constructed at Hanford in the USA during 1943 and 1944. Each generated 250 MW of heat and was cooled by water from the Columbia river; the moderator was graphite.

It is interesting to note that if fission had been discovered earlier in the earth's history, it would have been easier to construct reactors using natural uranium than it is now. This is because the U-235 content of natural uranium was higher in the past, the half-life of U-235 being much shorter than that of U-238. Moderation by ordinary water, instead of heavy water or graphite would have enabled a chain reaction to start, given a large enough mass of uranium. Indeed this is just what happened 1800 million years ago at Oklo in the Gabon in central Africa. At that time the proportion of U-235 in natural uranium was about 3%. An unusually rich ore body, in the presence of water which acted as a moderator, reached critically and a chain reaction continued, on and off, for a period of about a million years, using up about half the U-235 present. It is likely that other natural reactors have also occurred, but Oklo is the only one so far discovered. Studies of the subsequent movement of the fission products have given important information relevant to the disposal of radioactive wastes in deep geological formations (see p. 160).

The percentage of U-235 in uranium can be increased artificially (the uranium is said to be 'enriched') by making use of the small difference in physical properties between U-235 and U-238. Enrichment has the same effect as moderation: it increases the probability that a neutron will be absorbed by a U-235 nucleus and cause it to fission, rather than being absorbed by a U-238 nucleus and lost to the chain reaction. The first enrichment method to be applied to uranium was electromagnetic separation, using the fact that the trajectories of electrically charged atoms or 'ions' in a magnetic field depend on their ratio of charge to mass. The throughput turned out to be very small and the technique is now only used for the production of some special isotopes. The first method of enrichment used on a large scale was gaseous diffusion, which uses the fact that the rate at which a gas diffuses through a porous barrier depends on the reciprocal of the square root of the molecular weight of the gas. Lighter molecules travel faster than heavier ones and are therefore more likely to pass through the barrier. The diffusion process must be repeated many times because the fractional separation in a single stage is only about 1.002 and a typical plant consists of over 1000 stages arranged in cascades. The uranium is in the form of uranium hexafluoride, a highly corrosive gas, and the barriers are nickel tubes with pores roughly 0.01 micrometres in diameter (Leclerq 1986). The energy requirement for pumps and compressors is high and plants are generally powered by their own nuclear generating stations.

A less energy-intensive method of enrichment is the gas centrifuge in which the uranium hexafluoride gas is spun around at high speeds in a cylinder, with the heavier U-238 molecules tending to collect toward the outside of the cylinder. The energy requirements are about one-tenth of those of a diffusion plant for the same throughput and degree of enrichment.

A technique now attracting much attention is laser enrichment, which makes use of the differences in the absorption spectra of the two uranium isotopes. In principle, separation can be carried out in one step and the power requirements should be well below those of the diffusion and centrifuge methods. Two processes are being developed. In Molecular Laser Isotope Separation (MLIS) a laser is used to excite the U-235 hexafluoride molecules and a second laser is used to dissociate the excited molecules to form U-235 pentafluoride which is not a gas and is easily recovered. In Atomic Vapour Laser Isotope Separation (AVLIS) a uranium vapour is created by heating uranium metal in a vacuum with an electron beam. The U-235 atoms in the vapour are selectively ionized by a laser beam and the ions are collected electromagnetically (Leclerq 1986).

The majority of thermal reactors now use fuel enriched to about 3%, with ordinary water as a moderator. If enrichment is increased to about 10% a reactor can operate even without a moderator, because the chain reaction can be maintained by the fast neutrons that are emitted in the fission process. Such a reactor is called a fast reactor (strictly speaking a 'fast-neutron' reactor, in contrast to a 'slow-neutron' or thermal reactor). In order to achieve criticality and a nuclear explosion in a single lump of uranium metal, a much higher degree of enrichment and special technology is required to maintain the critical assembly for long enough to generate very high power before blowing itself apart.

We have seen how the absorption of neutrons by U-238 nuclei makes it difficult for a chain reaction to become self-sustaining: one needs either to moderate the neutrons to increase their chance of causing a fission or to increase the percentage of U-235 in the fuel. The consequences of this neutron absorption, however, can be highly beneficial. U-235 and U-238 nuclei both become unstable when struck by a neutron. But whereas the U-235 nucleus reacts by splitting into two roughly equal fragments, U-238 becomes U-239, which then emits a $\beta$ particle to become neptunium. This emits another $\beta$ particle to become plutonium—Pu-239. Pu-239, possibly created by Fermi but not detected by him, is a radioactive material with similar properties to U-235: it can readily sustain a chain reaction. Further absorption of neutrons in Pu-239 results in the production of Pu-241, which is also fissible. It was the realization that plutonium, like U-235, could be used for a weapon that led to the construction of the Hanford reactors.

The importance of plutonium for civil purposes is that it is the key to the utilization of U-238. It is this process, called 'breeding', which converts non-fissile U-238 into fissile plutonium. It enables the amount of energy that can be obtained from a given amount of uranium to be increased by about a hundredfold and turns uranium from a useful but limited source of energy into a virtually limitless one. This has major environmental consequences. The main effects are at the beginning and end of the fuel cycle: with a fully developed system of fast reactors breeding their own fissile fuel the quantity of uranium that has to be mined to produce a given amount of energy is greatly reduced, and the utilization of U-238 and plutonium, otherwise waste products, somewhat eases the problems of radioactive waste disposal. The key to the fast reactor is reprocessing—the chemical separation of plutonium and unused uranium from the remainder of the waste products.

## Radiation and its control

We have seen how the first distinguishing characteristic of nuclear power —its great concentration as an energy source—is almost entirely beneficial from the environmental point of view: mining, transport, land use, fuel storage, and quantities of waste are all very much less for nuclear than for fossil fuel stations. It is the second characteristic, the high level of radioactivity associated with the fission process, which constitutes the major potential impact of nuclear power on the environment and on health. Often industrial hazards, particularly those where the health effects do not appear until many years after exposure, are not recognized until the industry is well established and much damage has been done; asbestos is a prime example. A great deal of information on the risks of radiation had, however, been accumulated as a result of its use, particularly in medicine, well before nuclear fission was discovered.

Most of the key discoveries were made during the last few years of the nineteenth century (Saunders and Wade 1983). X-rays were discovered by Röntgen in 1895 and were rapidly applied: the first medical X-ray picture was used to locate shotgun pellets in a man's hand. The possible dangers were also recognized within a very short time. In 1896 Lord Lister, in a lecture to the British Association, said this of X-rays: 'If the skin is long exposed to their action it becomes very much irritated, affected with a sort of aggravating sun-burning. This suggests that transmission through the human body may not be altogether a matter of indifference to the internal organs.' (Posner 1970). In the very early years X-rays were actually measured in terms of the erythema dose, an empirical biological quantity which was the dose needed to produce the erythema, or skin irritation described by Lister.

In 1896 Becquerel found that pitchblende, an ore of uranium, caused the darkening of a photographic plate; he had discovered radioactivity. Pierre and Marie Curie extracted a small amount of radium from pitchblende later that year. The hazards of radium were soon appreciated. Rutherford visited the Curies in 1903 and noted 'We could not help but observe that the hands of Professor Curie were in a very inflamed and painful state due to exposure to radium rays' (Eve 1939). Rutherford identified two types of radiation from pitchblende, with different penetrating power; he called these alpha and beta rays. Villard identified a third component, which he called gamma rays. Rutherford showed that alpha rays consisted of helium nuclei, Becquerel identified beta rays as electrons, and gamma rays were later shown to be electromagnetic radiation similar to, but more penetrating than X-rays. The neutron was discovered much later, by Chadwick in 1932.

Although the existence of radiation was only appreciated as a result of these discoveries, it was not, of course, a new phenomenon. The universe has been radioactive since its formation and all life-forms have evolved in a radioactive environment—indeed radiation almost certainly played a part in evolution through the mutations that it can produce. We are all continuously bombarded by cosmic radiation from the sun and from outside the solar system, from our naturally radioactive surroundings, and from naturally occurring radionuclides that we breathe, eat, and drink. Every hour about 500 000 cosmic ray particles pass through us, about 30 000 radioactive atoms from the air we breathe disintegrate in our lungs, some 15 million potassium-40 atoms disintegrate somewhere in our bodies and 200 million gamma rays pass through us from our surroundings (NRPB 1986).

All of this radiation and any additional amounts we receive from artificial sources can in principle harm us. The basic process by which it does so is common to all types of radiation: energy is transferred from the radiation to the tissue through which it passes. The energy lost by the radiation appears as excitation and ionization energy, mainly in the water molecules which lie along the track of the radiation (stritcly speaking the radiation we are discussing should, for this reason, be called ionizing radiation; other forms of radiation such as sunlight and radio waves can also be harmful in some circumstances but do not cause ionization). This initial process is rapidly followed by molecular reorganizations. All molecules can be altered by radiation in this way; the interactions are highly complex but in living matter the DNA macromolecules that carry the genetic information required for the development and division of cells are probably the most critical targets.

The fact that we can withstand the constant bombardment of radiation from natural sources without apparent ill-effect suggests that virtually all the damage at the molecular level is either unimportant or readily repaired by the body's natural repair mechanisms. There are, however, two ways in which the damage can be important: it can result in cell death and in cell transformation. Dead cells are normally absorbed or rejected by the organism. However, if a sufficient number of cells are killed, function will be affected and the organism may die. Cell transformations or mutations do not necessarily lead to any harmful effects; indeed, very large numbers of such cellular changes occur normally during the lifetime of any organism. They may, however, result in a cancer or, in the case of cells responsible for reproduction, in genetic damage in later generations.

The quantitative description of the effects of radiation depends on the availability of a system of units. The radioactivity (often shortened to 'activity') of a given amount of material is measured simply by the rate at which the radionuclei present transform spontaneously into other nuclei, with the emission of radiation. Originally activity was measured in curies (Ci); one curie equals $3.7 \times 10^{10}$ disintegrations per second. The unit now used is the becquerel (Bq); one becquerel equals one disintegration per second.

Radiation is measured in terms of the energy absorbed (or 'absorbed dose') in the matter through which it passes. The original unit was the rad, with 1 rad equalling 100 ergs of energy absorbed per gram of matter or 0.01 joules per kilogram. The unit now used in the gray (Gy), with 1 gray equalling 1 joule per kilogram (1 Gy=100 rad).

While energy absorption is a reasonable measure for radiation itself, it is not a good measure of its biological effects. This is because the ability of radiation to harm biological matter depends on the details of the energy transfer mechanism as well as on the amount absorbed. The way in which radiation loses energy depends on the type of radiation involved and on its energy. Heavy particles such as neutrons and alpha particles produce more ionization in a given distance than beta particles or gamma rays. Moreover, the amount of energy absorbed varies continuously along the track of the particle as it slows down and tissue along a particle track is also irradiated by secondary particles resulting from primary ionizations in nearby tracks. Thus while the absorbed dose is in principle relatively easy to assess, a complete description of the total effect of a given exposure to radiation requires a precise knowledge of the type and energy of the radiation and the irradiation conditions.

For the purposes of radiological protection it proves sufficiently accurate to use a quality factor (Q) to convert the energy deposition measured by the absorbed dose (in rads or grays) into a 'dose-equivalent' which is a measure of the biological effectiveness of the radiation. The unit of dose equivalent first introduced was the rem, with 1 rem again equalling 100 ergs of energy absorbed per gram of matter, or 0.01 joules per kilogram, and

$$1 \text{ rem} = 1 \text{ rad} \times Q.$$

The unit now used is the sievert (Sv), with one sievert equalling 100 rem and

$$1 \text{ Sv} = 1 \text{ Gy} \times Q.$$

The quality factors for alpha particles and neutrons are twenty and ten respectively; for beta particles, X-rays and gamma rays Q=1. So one gray of alpha particles, for example, gives a dose-equivalent of 20 sieverts while one gray of gamma rays gives a dose-equivalent of one sievert.

The biological effects of radiation depend on the characteristics of the target tissue as well as on the type of radiation. Cells that are dividing frequently, such as the cells in blood-forming tissue in bone marrow, are more sensitive than cells that only divide occasionally, such as cells in fat. Metabolic factors, such as the concentration of oxygen in the irradiated volume, are also important. Finally, the age and sex and to some extent the health of the organism also influence the extent to which radiation can do harm.

Radiation can interact with tissue either directly, from an external source, or indirectly, following the intake of radioactive material. The relationship betwen an exposure to radiation or an intake of radioactive material and any subsequent biological effects can therefore be very complex. For the purposes of radiological protection a simplified approach is used. The way in which radioactive material taken into the body is metabolised and irradiates various organs is modelled using an average 'Reference Man' (ICRP 1975). The different sensitivities of different parts of the body are allowed for by the use of weighting factors; these are given in Table 3.1. The 'weighted dose equivalent', also called the 'effective dose-equivalent' is also expressed in sieverts. A dose-equivalent of one sievert to the red bone marrow, for example, gives an effective dose-equivalent of 0.12 Sv while the same dose-equivalent to the thyroid gives an effective dose-equivalent of 0.03 Sv.

**Table 3.1**
Weighting factors ($W_T$)

| Tissue | $W_T$ |
|---|---|
| Gonads | 0.25 |
| Breast | 0.15 |
| Red bone marrow | 0.12 |
| Lung | 0.12 |
| Thyroid | 0.03 |
| Bone surfaces | 0.03 |
| Remainder | 0.30 |

*Source*: ICRP 1976

For direct external radiation, the damage can only occur while the organ is being exposed to the radiation. For ingested and inhaled radioactivity, however, the duration of the exposure will depend on the rate at which the material is metabolized and excreted and on its half-life, the time taken for its radioactivity to fall by one half. Some long-lived radionuclides that are not excreted can irradiate organs during the rest of the life of the organism. The total dose resulting from the intake of radioactive material, weighted according to the part of the body which the material irradiates, is called the 'effective dose-equivalent commitment'; this, not surprisingly, is commonly abbreviated to 'dose' and measured in sieverts.

Finally, a source of radiation or a quantity of radioactive material can irradiate a large group of people or a whole population, and it is sometimes necessary to assess the effects on such a population. For this purpose we use the complicated but precise term 'collective effective dose-equivalent 'commitment', commonly abbreviated to 'collective dose' and measured in man-sieverts, or, by some, in person-sieverts.

The sievert is the measure used in those radiological protection applications where the principal effects are due to cell transformation. Where cell death predominates, generally at very high doses, the biological effects of different types of radiation may vary in different ways and the gray remains the appropriate unit.

The sievert and the gray are large units; natural background radiation, for example, results in an average dose in the UK of about two thousandths of a sievert a year (2 mSv).

## Early effects

We have seen that radiation can lead to cell killing and cell transformation. Cell killing is medically significant only if a sufficiently large dose of radiation is received in a sufficiently short time. For humans, a dose of 10 Gy or more delivered to the whole or a substantial part of the body within a few minutes is almost invariably fatal. A single dose of about 4 Gy will result in about a one in two chance of death in the absence of medical treatment. The same dose spread over many weeks, however, would probably have no effect because the rate at which cells would be killed would be low enough to enable them to be replaced by natural processes. Other effects of high exposures are skin burns, loss of hair, radiation sickness, reduction in fertility, and cataract of the eye. All these effects occur within hours, days, or weeks of the radiation exposure and are therefore called 'early effects'. Cataract of the eye is an exception, occurring, if at all, after many years. The ability of radiation to kill cells is, of course, the basis of radiotherapy where localized doses of tens of grays are used to treat cancers and other growths.

Early effects and cataract induction are characterized by a threshold, below which no effects are observed. Above this level the severity of the effect increases as the dose increases. The threshold is about 1 Gy, except for the embryo where at certain stages of development only a few cells may be involved in important stages of growth and a much smaller dose can result in the destruction of these cells and lead to abortion or serious malformation.

Radiation doses that are insufficient to cause these early effects may, as a result of cell transformations, cause damage that does not show up until many years after the radiation exposure has occurred. The main types of 'delayed effects' are cancers in the exposed individual and genetic defects in his or her descendants. Unlike the early effects, which inevitably occur if a sufficiently large dose is received and for which the size of the dose governs the severity of the effect, the delayed effects only occur in a small proportion of those irradiated, in an apparently random way, and the size of the dose governs the probability of the effect occurring, but not the severity of the effect itself. In this way the delayed effects of radiation are similar to those of chemical carcinogens, asbestos, or tobacco. They are called 'stochastic' effects because they are governed by the rules of chance. The early effects are called 'non-stochastic'.

### Radiation-induced cancer

While the first indications of the damaging effects of radiation were skin burns due to cell death, it was soon realized that Lister's warning 'that the transmission of radiation through the body was not altogether a matter of indifference to the internal organs' was justified. Cancers were observed both in the areas in which acute damage had occurred and in other parts of the body that had been exposed. Early radiologists took few precautions to avoid exposure and began to experience higher than expected cancer rates. Cancers resulted from the use of radiation to treat non-malignant conditions. It is difficult to derive any quantitative estimate of risk from these early observations because there was essentially no measure of radiation dose.

There is now a limited but reasonably consistent body of evidence on which quantitative estimates of the risks of cancer induction by doses of radiation of the order of a few tenths of a sievert and above can be made.

The information comes from three classes of exposure: occupational, medical, and military. The most comprehensive reviews of the evidence are those carried out by the United Nations Scientific Committee on the Effects of Atomic Radiation (UNSCEAR, 1988). The evidence is based on a total of only around two thousand cancer deaths that can definitely be ascribed to radiation since it was discovered and for which reasonable estimates of radiation dose can be made—a surprisingly low figure in the light of the common dread of radiation, specially since it includes the cancer deaths at Hiroshima and Nagasaki. The number is also small when compared with the total numbers in the exposed populations; Table 3.2 shows that the highest proportional excess occurred among the radium luminizers: people painting figures on the dials of watches and clocks to make them visible in the dark in the early days before the risks were appreciated used to lick the brushes to give them a fine point. Some radium was inevitably swallowed and bone cancers were observed to occur some years later. The other main sources of evidence from occupational exposure were radiologists, and uranium and other hard-rock miners working in inadequately ventilated mines. It is not known just how many died as a result of the early medical uses of radiation, although there is a monument in Hamburg to the early 'martyrs of radiation' which records 169 deaths among radiologists alone (Pochin 1983). No quantitative information can be derived from these tragedies because very little is known about the total radiation doses received. It was these deaths, however, together with the serious skin injuries among early workers that led to the introduction of radiological protection measures from 1913 onwards (see page 95).

The earliest record of deaths from occupational exposure to radiation, although the cause was not, of course, known at the time, was the appallingly high incidence of 'black lung' among the metalliferous miners of Saxony from the fifteenth century onwards. This is now known to have been lung cancer caused by the high concentrations of the radioactive gas radon in the inadequately ventilated tunnels. The radon risk to uranium miners today is discussed later.

The most precise estimates of the numbers of cancers that can result from radiation exposure comes from the use of radiation and radioactive materials in medical therapy, and, in some instances, diagnosis. Although there is often some uncertainty about the size of the irradiated volume and the precise radiation doses received, the availability of medical records giving details of the treatment and subsequent health of the patients and the existence of control groups suffering from the same condition but not treated with radiation give reasonable confidence in the risk estimates derived for these irradiation conditions and for the cancers involved. It does not follow that the results can be extrapolated to other situations where the irradiation conditions may be very different.

The largest study, in terms of the size of the exposed population, is that of the survivors of the Hiroshima and Nagasaki bombs. The uncertainty here lies in the sizes of the doses received. This depended both on the quantities of different types of radiation produced by the bombs and on the position of the individuals at the precise time of the explosions, including such factors as shielding by buildings. The figures given in Table 3.2 are derived from a recent reassessment of the data (Shimizu *et al.* 1987).

There are a number of differences between the occupational, medical, and military exposures. Those occupationally exposed can be assumed to be reasonably healthy, indeed in making comparisons between their subsequent health records and those of the general population some allowance may have to be made for the 'healthy worker effect' — unhealthy people are less likely to be employed. Workers' exposure is generally at a low rate and over a long period. Medical exposures are also to a selected population—those needing treatment. The irradiation is often highly localized and delivered in large doses, sometimes repeated over periods of some weeks. The exposures at Hiroshima and Nagasaki were very intense and brief. The effects of the bombings were so severe that the very fact of survival may constitute an element of selection.

In spite of these differences there is a reasonable consistency in the estimates of cancer risk that can be made and this gives some confidence in the application of these estimates to situations that are again different.

**Table 3.2**
Principal epidemiological evidence for cancer induction by radiation

| Source of exposure | Principal radiation | Numbers irradiated | Observed excess numbers of deaths (or Diagnoses) from certain malignancies | | | | |
| --- | --- | --- | --- | --- | --- | --- | --- |
| | | | Leukaemia | Thyroid[+] | Bone | Lung | Breast |
| OCCUPATIONAL | | | | | | | |
| Radium Luminizers | α | 780 | | | 51 | | |
| Miners — US, uranium | α | 3,366 | | | | 58 | |
| — Czech, uranium | α | | | | | 38 | |
| — Swedish, non-uranium | α | 7,500 | | | | 20 | |
| — Newfoundland, fluorite | α | 1,100 | | | | 61 | |
| MEDICAL | | | | | | | |
| Ankylosing Spondylitis | X | 14,109 | 24 | | 3 | 3 | |

| | | | | | |
|---|---|---|---|---|---|
| Metropathia Haemorrhagia | X | 2,068 | 5 | | |
| Thymic Enlargement | X | 2,876 | 4 | 1 (23) | |
| Tinea Capitis | X | 10,902 | | (10) | 1 |
| Tinea Capitis | X | 2,892 | 5 | | |
| Radium-224 treatment | α | 815 | | | 47 |
| TB Fluoroscopy — Massachusetts | X | 1,047 | | | 18 |
| TB Fluoroscopy — Nova Scotia | X | 326 | | | 28 |
| Mastitis | X | 606 | | | 20 |
| Benign diseases — Sweden | X | 1,115 | | | 93 |
| Thorotrast | α | 4,594 | 54 | | |
| Tonsil, Nasopharynx Group | X | 859 | | (57) | |
| **MILITARY** | | | | | |
| Hiroshima and Nagasaki | γ | 41,719* | 74 | | 318 (all cancers except leukaemia) |

*Exposure doses above 0.01 Gy, the 0–0.01 Gy group is used as a control, data to end 1985.
+ Most thyroid cancers are cured. The figures in brackets give the number of cases diagnosed.

Estimates have been made by the National Research Council Committee on the Biological Effects of Ionizing Radiation—the BEIR committee (BEIR 1980) as well as by UNSCEAR. The total risk of fatal cancer, averaged over all ages and both sexes, is now estimated to be between 4 and 11 cases for 100 people all exposed to a dose of one sievert, if that dose is delivered at a high rate (UNSCEAR 1988). The risk of exposure to lower doses and at lower dose rates is discussed below.

## Genetic effects

If radiation damages the DNA in a cell in the reproductive organs, and if that cell is the one actually involved in the creation of a new individual, the damage can be passed on to that new individual and may show up as an abnormality.

The DNA molecule consists of a linked double helix, the backbone of sugar and phosphate molecules serving as a framework to hold in place the four different types of bases whose sequence contains the genetic information. There are two ways in which heritable changes in the DNA can occur. The first is the removal, replacement, alteration, or interchange of one or more of the base molecules, called a gene or point mutation, and the second is a break of one or both backbones of the double helix, called a chromosomal mutation. A single ionization or 'hit' by radiation can cause either of these types of damage.

In contrast to the case of radiation-induced cancers, there is very little information on radiation-induced genetic effects in humans. Some studies carried out on populations living in areas of high natural background radiation suggested that there was a correlation between one particular defect, Down syndrome, and radiation levels but subsequent analysis has thrown doubt on this conclusion (Rose 1982). There is some association between maternal X-irradiation and spontaneous abortion of chromosomally fetuses, but such fetuses are usually not viable and are lost early in pregnancy. There is evidence from radiation workers that chromosomal mutations can be induced in blood cells, but there is no evidence of any associated detrimental effect, or of any genetic effect. No genetic defects that can unequivocally be ascribed to the effects of radiation have ever been found in the offspring of irradiated humans. The most extensive studies are those carried out on the children of the survivors of the Hiroshima and Nagasaki bombs. Studies of still-births, infant mortality, low birth weight, frequency of congenital malformations, sex ratio, cancer incidence and subsequent growth and development show no statistically significant differences between the children conceived by exposed parents and suitable control groups. However, the findings, while not statistically significant, are generally in the direction that would be expected were there to be an effect due to radiation (UNSCEAR, 1982).

Estimates of the incidence of radiation-induced genetic effects in humans have therefore to be based on extrapolation from animal experiments. Almost all the evidence comes from large-scale studies of mice and although there are some basic similarities in the genetic material a great many assumptions have to be made in extrapolating from mouse to man. There are two ways in which the effects in man are estimated. In the first, radiation-induced mutation rates derived from experiments on mice are compared with spontaneous mutation rates derived from surveys of the natural incidence of hereditary disorders in man, and the dose of radiation required to double the spontaneous rate in man is calculated. This is known as the 'doubling dose' method. In the second, data on radiation-induced mutation rates in mice are used to estimate directly the number of mutations that will occur in man as a result of a given amount of radiation. This is known as the 'direct' method.

Both methods involve considerable uncertainties. In the doubling dose method it is assumed that similar doubling doses apply to mice and men, and that the sensitivity of a gene to a spontaneous mutation and to a radiation-induced mutation is the same. Doubling doses derived from mouse experiments range from 0.4–2.6 Sv, and an average value of 1 Sv is used in most estimates of effects in man. The Hiroshima and Nagasaki studies suggest that the hazard estimated by extrapolation from animal data using a doubling dose of 1 Sv is unlikely to be significantly underestimated.

The doubling dose method expresses radiation-induced genetic disorders in terms of naturally occurring disorders; the latter represent an equilibrium between new mutations and mutations that are lost from the population by death or failure to reproduce. A doubling dose thus has to be maintained for several generations (5 to 10) before equilibrium is reached and the number of radiation-induced disorders equals the number of naturally occurring disorders (BEIR 1980). The first generation incidence is between one-fifth and one-tenth of the equilibrium incidence.

The direct method of estimating the genetic effects of a dose of radiation uses data on specific types of radiation-induced defect in mice to estimate directly the number of defects that will occur in man as a result of a given amount of radiation. As with the doubling dose method, the major uncertainty stems from the assumption that mouse data are relevant to man. In addition, calculation of the rate of induction of all defects in man from data relating to only one kind of defect in mice (skeletal abnormalities) required knowledge of the proportion of defects in man whose main effect is in the skeleton. In spite of the uncertainties involved there is reasonable agreement between the predictions of the direct method and the doubling dose method.

As with radiation-induced cancers, there is reasonable agreement between the estimates that have been made, and the consensus view is that the total number of genetic defects, spread over all subsequent generations, resulting from a given dose of radiation to a population, is of the same order as, and probably somewhat below, the total number of fatal cancers that would occur in the exposed population itself as a result of that radiation.

### Radiation and expectation of life

It has been known since 1939 that animals exposed to large doses of radiation over and above those received from the natural background have a shorter life span than control animals not receiving the additional radiation. For many years it was thought that this may be due to some non-specific ageing effect in addition to the specific induction of cancers. The subject was extensively reviewed in UNSCEAR (1982). Information is available from animal studies and from epidemiological studies of human populations, such as those that form the basis for estimates of the risk of radiation-induced cancer. UNSCEAR concluded that there was no firm evidence for any life-shortening, either in animals or in man, except that caused by the induction of cancer.

Evidence for the reverse process—an increase in life expectation as a result of irradiation—is less well-known but extensive (Luckey 1980). Life extension resulting from small additional doses of radiation has been found in a very wide range of animal and plant species. This finding is not inconsistent with the view that any dose of radiation, however small, can result in damage to DNA which can, in turn, result in cancer in organisms susceptible to that disease. The mechanism that has been proposed is 'radiation hormesis', hormesis being the beneficial stimulation of a biological system by a small dose of an agent that can be harmful at high doses. The hypothesis is that natural repair mechanisms are 'triggered' by small doses of radiation so that the organism is better able to resist subsequent attack, by radiation or any other agent, leading to improvement in overall health and expectation of life. While the phenomenon of radiation hormesis indubitably exists and may find application in plant breeding, it is generally assumed that for humans the harmful effects of radiation exceed any possible beneficial effects to health, with the obvious exception of medical uses of radiation and radioactive materials.

### Effects of low dose levels

The critical evidence on the effects of radiation in man is based on observations of the effects at doses of around half a sievert and above. There are some indications that occasional malignancies may in certain circumstances be induced by single doses of about one tenth of a sievert but these indications do not provide an adequate basis for calculating the risks of a similar dose spread over a long period. And it is just at such exposure levels and below that estimates need to be made in order to assess the environmental impact of nuclear electricity generation. The maximum dose received in a whole year by any member of the public in the UK as a result of the activities of the nuclear industry is now somewhat below one two-thousandth of a sievert (0.5 millisievert). Such a dose is received by only a small group of people who eat unusually large amounts of fish and other seafoods caught in the Irish Sea near the vicinity of the Sellafield reprocessing plant. Typical doses to people living within a few kilometres of nuclear power stations are around one hundredth of a millisievert (10 microsieverts) a year and the average annual U.K. dose from all stages of the nuclear fuel cycle is less than one thousandth of a millisievert (1 microsievert). What, if any, are the risks associated with such levels of exposure?

It is unlikely that a direct answer to this question will ever be found. The first problem is that of statistics. If the risk per unit of radiation exposure is the same at low levels as at the levels where direct evidence is available, to detect the effects of a dose of a few millisieverts over and above natural background with reasonable statistical accuracy one would need to study a population of about one million. One would also need a similarly large control population not exposed to the additional dose, and the study would have to continue over the many decades required for all the cancers to appear. The only possibility of carrying out such studies is to find large populations that have been exposed to substantially more than the average from natural background radiation. Attempts have been made to find correlations between natural cancer incidence and background radiation levels and no statistically significant effects have been found. Indeed areas with high levels of background radiation sometimes have significantly lower numbers of cancer deaths than areas with lower background radiation levels. In the US it is found that states with a high level of background radiation, like Colorado, have a significantly lower cancer death rate than the eastern seaboard states where radiation levels are lower by about a factor of two (Frigerio et al. 1973). A recent study in India has shown that cancer death and cancer incidence rates in five cities decrease with increasing natural background radiation levels (Nambi 1987). A similar result was found in China, where detailed studies were carried out of cancer mortality in two areas; in one the natural background radiation level was over twice that in the other, yet the cancer mortality rate was somewhat lower (Pochin 1989). These results do not mean that radiation is necessarily good for you, although the possibility of radiation hormesis cannot be ruled out, but strongly suggest that these background levels are not a major cause of cancer.

The second problem is that cancer is a very common form of death, accounting for about one in four of all deaths in the UK, for example, and moreover that cancer death rates vary by large amounts in a way not generally understood with such factors as geography, life style, social class, and eating habits.

Since the risk of cancer being induced by low levels of radiation will probably never be derived from direct observation one must use indirect methods. The first is through the development of fundamental knowledge in radiobiology and the second is by observing the relationship between radiation dose and cancer incidence in the range of doses in which cancers can reliably be linked to radiation and assuming that a similar relationship holds for lower doses.

Our understanding of the ways in which radiation can damage DNA and of the ways in which such damage can be repaired is consistent with a different form of relationship between radiation dose and effect for radiation for which the quality factor Q equals one, such as beta particles and gamma rays, and radiation for which Q is large, such as alpha particles (Pochin 1983). Much of the damage caused by low doses of beta particles and gamma rays is repaired, typically within minutes, and the risk associated with such doses is less than would be inferred from observations at high doses. Alpha particles, however, appear to produce damage that is harder to repair, and the risk would be expected to be more nearly proportional to the radiation dose. This picture is generally confirmed by studies of the ways effects in cells and cancer incidence in animals actually varies with dose for the different types of radiation (UNSCEAR 1986).

Inevitably, in the absence of direct evidence, there is a wide spectrum of views on the effects of radiation at levels well below those at which these studies have been carried out. At one extreme, there are those who hold that low doses are proportionally more damaging than high doses. At the other there are those who quote the large body of evidence relating to radiation hormesis, showing an increase in life expectancy resulting from low-level irradiation. The BEIR (1980) and UNSCEAR (1988) reports both comment on the applicability of their risk estimates to low levels. The BEIR report states that the effects of background radiation are unknown; it stresses that the estimates given relate only to doses ten or more times greater than annual background doses and probably overestimate the risk from lower doses. The UNSCEAR report states that there is 'a need for a reduction factor to modify the risks (derived from evidence at high dose rates) for low doses and low dose rates. The Committee considered that such a factor varies very widely with individual tumour type and with dose rate range. However, an appropriate range to be applied to total risk for low dose and low dose rate should lie between 2 and 10'.

## Radiological protection

We have seen how rapidly the discovery of radiation was followed by a realization of at least some of its hazards. While individual workers probably developed their own methods of reducing their exposure, the first formal step appears to have been taken by the Deutsche Roentgen Gesellschaft which in 1913 issued a leaflet making specific recommendations on the thicknesses of lead screening to be used when working with X-rays. In Britain the Roentgen Society in 1915 agreed 'That in view of the recent large increase in the number of X-ray installations, this Society considers it a matter of the greatest importance that the personal safety of the operators should be secured by the universal adoption of stringent rules' The extensive medical use of X-rays during the 1914–18 war, often in the difficult conditions of field hospitals, led to a further realization of the need to protect those involved.

The International Congress of Radiology met for the first time in London in 1925 and at the second meeting in Stockholm in 1928 it was agreed to form an International Committee on X-ray and Radium Protection, later renamed the International Commission on Radiological Protection (ICRP). The earliest recommendations of the ICRP included minimum thicknesses of lead shielding for X-ray installations and radium sources and referred to the general need for good housekeeping and working conditions, but did not give any numerical guidance on the limitation of exposure. At the time it was thought that radiation-induced cancers only occurred in tissue that had been visibly damaged by radiation. Since such damage did not occur below a certain level of exposure this implied that there was a threshold below which no cancers would result. In 1925 Mutscheller in Germany and Sievert in Sweden independently suggested that an acceptable dose for people working with radiation was one tenth of an erythema dose (the dose required to produce a reddening of the skin) per year (Stone 1952). This corresponds to about 0.3–0.7 Sv per year.

ICRP first made recommendations on a level of exposure that could be tolerated by a person in normal health operating under satisfactory working conditions in 1934; these corresponded to 2 mSv per day or around 0.7 Sv per year, comparable to the earlier recommendations of Mutscheller and Sievert. Even in these early days of radiological protection, however, the recommendations include warnings that the operator 'should on no account expose himself unnecessarily' and 'should place himself as remote as practical from the X-ray tube', ideas very close to today's recommendations.

The first doubts about the existence of a threshold emerged during the 1930s from experiments on radiation-induced mutations: Muller had found in 1927 that mutation rates in fruit flies (Drosophila) could be accelerated by irradiation with X-rays (Muller 1927). In the early 1940s the Advisory Committee on X-ray and Radium Protection in the US discussed the possibility of reducing the recommended tolerance dose by a factor of ten to take into account the possibility of genetic damage.

A large amount of information accumulated during the Second World War and in the following years as a result of the massive programme to develop nuclear weapons and their use at Hiroshima and Nagasaki. ICRP was re-formed in 1950 and recommended a maximum permissible dose corresponding to 0.15 Sv per year, about one-fifth of the pre-war level. 'Permissible dose' replaced 'tolerance dose' to reflect the doubts about the existence of a threshold and it was recommended that 'every effort be made to reduce exposure to all types of ionizing radiations to the lowest possible level'.

During the 1950s information on the delayed effects of the Hiroshima and Nagasaki weapons and the rapid growth in the use of sources of radiation in industry and medicine led to an important change in protection philosophy. ICRP's 1958 recommendations were based on the abandonment of the idea of a threshold, although it was stressed that this was an assumption not supported by any evidence. Since it was assumed that any dose, however small, carried some risk it was no longer possible to base a system of protection on the idea of absolute safety provided certain limits were adhered to. It became necessary to make judgements about the acceptability of the risks that were assumed to exist.

The aim of radiation protection, as set out in ICRP 26, 1976 is to 'prevent detrimental non-stochastic effects (i.e. those associated with cell killing and exhibiting a threshold) and to limit the probability of stochastic effects (i.e. cancers and genetic damage, assumed to be linearly related to dose, without a threshold) to levels deemed to be acceptable.' ICRP's current approach is formally expressed as three principles:

1. All practices involving radiation exposure should show a net positive benefit (justification).

2. Radiation doses should be as low as reasonably achievable, economic and social factors being taken into account (optimization).

3. All radiation exposures should be within the recommended dose limits.

The first principle results from the assumption that any exposure to radiation, however small, may cause some harm and should therefore not be allowed without good reason. It has resulted in the banning of trivial uses of radioactive material, such as in toys and jewellery. More importantly, it calls for care in situations where an application is of clear benefit but carries some risk; a good example is X-ray screening for breast cancer where the number of cancers prevented should not be exceeded by the number induced by the examination itself. This requirement is probably satisfied by modern radiological techniques but may not always have been in the past. The principle is more difficult to apply when the benefits are general but the risks are concentrated on a few, for example those living near a particular nuclear plant. This problem has to be faced in many fields such as the siting of airports, motorways, and chemical plants and is related to the main current problem in radioactive waste disposal, the so-called NIMBY (not in my back yard) syndrome. The question was also addressed by Layfield folowing the Inquiry into the siting of Britain's first pressurised water reactor at Sizewell in Suffolk (see p. 115).

The second principle, commonly known as ALARA (as low as reasonably achievable), presents one of the current challenges in radiological protection. The requirement is straightforward: where the cost of reducing an exposure to radiation is less than the resulting harm arising from that exposure, then the exposure should be reduced. The difficulty is in expressing the harm, which may consist of an increased risk of cancer or genetic damage to an individual, a group or a large population, in monetary terms. While setting a monetary value on death or disease may appear repugnant, it often needs to be done, either explicitly or implicitly, when taking decisions about the resources to be devoted to health care or accident prevention in general. The amounts of money that are actually spent directly on saving a life or on reducing risks vary enormously. At one extreme, a figure of £120 million per life saved is associated with the decision by British Nuclear Fuels plc in 1984 to reduce further the discharges from its reprocessing plant at Sellafield to the Irish Sea (Avery 1984). At the other, famine relief organizations can save the life of a baby facing death from malnutrition for a few pounds.

In the UK the National Radiological Protection Board (NRPB) is responsible for advising the government on radiological protection. It has proposed monetary values for radiation exposure that depend on the level of dose on the basis that more should be spent on reducing larger doses that carry more risk than on reducing small doses that carry less. The values proposed by NRPB are £3000 per man-sievert (corresponding to £150 000 per life saved) at the lowest doses and ten times that (corresponding to £1.5 million per life saved) at doses approaching the dose limit.

An important factor in considering the ALARA principle is that of the significance of very small doses that may be received by very large populations or spread over long periods. In terms of collective dose the exposure may be large but in terms of individual dose, the risk may be almost vanishingly small. The NRPB guidelines that reduce the monetary value per unit of dose as the dose itself is reduced go some way to tackling this problem, but they allocate the same 'value' to the reduction of one man-sievert of collective dose if it is distributed between ten million people, each receiving one ten-millionth of a sievert (0.1 $\mu$Sv) as between a hundred thousand, each receiving one hundred-thousandth of a sievert (10 $\mu$Sv). It would seem more reasonable to ignore any exposure that does not result in a significant risk to any individual at any time and NRPB's latest recommendations (NRPB 1985) go some way to reflecting this view.

The third principle, dose limits, used to be the main requirement when it was believed that keeping doses below a threshold level was all that was needed to ensure safety. It has now been relegated to a protection back-stop to ensure that no individuals are exposed to unacceptable risks. ICRP's current recommended dose limit of 50 mSv per year for radiation workers is based on the assumption that the need to keep within this limit and the observation of the ALARA principle will in practice result in average occupational doses of about 5 mSv per year, or two or three times the background dose. NRPB has issued guidance on the implications of the latest Hiroshima and Nagasaki risk estimates (see p. 87) and is recommending that 'occupational workers' exposure should be so controlled as not to exceed an average effective dose equivalent of 15 mSv per year' (NRPB 1987b). The average doses actually received by radiation workers in Britain are shown in Table 3.3 (Hughes 1988). Based on the current NRPB estimates (NRPB 1988), a worker receiving a dose of 2 mSv every year over a working life of 40 years would increase his chance of eventually dying of cancer from about 0.200, the average UK figure, to between 0.201 and 0.204, an increase of between 0.5% and 2%. This is comparable to the risks of death from occupational accidents in industries generally described as safe.

A number of attempts have been made to establish whether radiation workers actually suffer higher cancer rates as a result of their employment. While this was undoubtedly the case in the early years of the century, as already described, no consistent pattern has emerged from recent studies. Several studies purporting to find statistically significant links between occupational exposure and enhanced overall cancer rates have been refuted by subsequent analysis (BEIR 1980).

More recently, mortality studies of UKAEA and Sellafield workers have been published (Beral et al. 1985, Smith and Douglas 1986). Both show the workers to have a lower than average overall cancer death rate. When examining specific cancer sites it must be expected that some statistically significant relationships between mortality and dose will be observed by chance (Beral et al. 1985). In these studies, some cancers are indeed found to occur more frequently than expected and some, actually the majority, to occur less often than expected. Inevitably, the former have attracted more attention, in particular prostate cancer in the UKAEA study and multiple myeloma in a similar study of radiation workers at the Hanford plant in the

**Table 3.3**

Occupational radiation exposure (UK)

| Type of work | Total number of workers | Average annual dose (mSv) |
|---|---|---|
| Nuclear | | |
| Fuel fabrication | 3,404 | 2.8 |
| Fuel enrichment | 988 | 0.2 |
| Fuel reprocessing | 9,407 | 4.2 |
| Power stations | 22,971 | 0.8 |
| Research establishments | 9,195 | 2.7 |
| All nuclear industry | 45,965 | 2.0 |
| Defence | 16,860 | 1.0 |
| General industry | 21,000 | 0.6 |
| Tertiary education | 13,000 | 0.1 |
| Health | | |
| Medical | 40,000 | 0.2 |
| Dental | 20,000 | 0.1 |
| Veterinary | 4,000 | 0.1 |
| Natural | | |
| Coal mines | 81,500 | 1.2 |
| Non-coal mines | 2,000 | 14.0 |
| Aircraft crew | 20,000 | 2.0 |
| Total (rounded) | 260,000 | 1.1 |

USA (Tolley *et al.* 1983). When the data from the UKAEA and the Hanford Studies are combined, however, the results for these two cancers are no longer significantly different from zero, suggesting that these observed excesses in the individual studies were due to chance (Darby *et al.* 1985).

ICRP recommend a limit for members of the public, lower than for workers, on the basis of comparisons with everyday risks:

The acceptable level of risk for stochastic phenomena for members of the general public may be inferred from consideration of risks that an individual can modify to only a small degree and which, like radiation safety, may be regulated by national ordinance. An example of such risks is that of using public transport. From a review of available information related to risks regularly accepted in every day life it can be concluded that the level of acceptability for fatal risks to the general public is an order of magnitude lower than for occupational risks (ICRP 1976).

The public dose limit is currently set at 1 mSv per year, although 5 mSv is permissible in any particular year provided that the average annual exposure over a lifetime does not exceed 1 mSv. The dose limit is set to protect those most likely to be at risk as a result of a particular operation, called the critical group. NRPB's latest guidance recommends that 'critical group doses from effluent discharges from nuclear installations should be so controlled as not to exceed an effective dose equivalent of 0.5 mSv per year for a single site' (NRPB 1987b).

In order to ensure that the exposure of the critical group remains within the limit it is necessary to understand in detail the routes or 'pathways' by which radioactive materials released into the environment reach people. The subject is a complex one involving a combination of mathematical modelling and environmental measurements. Generally it is found that one or two pathways dominate and these are called the 'critical pathways'. Discharges of radioactive material are only allowed if the authorizing departments (the Department of the Environment and the Ministry of Agriculture, Fisheries, and Food in England and the Secretaries of State in Scotland and Wales) are satisfied both that the ICRP limit will not be exceeded, even for the critical group, and that the discharges are as low as reasonably achievable. Regular assessments carried out by the regulatory authorities have shown that critical group doses in Britain have never significantly exceeded ICRP limits. Proper protection of the critical group, of course, ensures that average doses are very much lower and in Britain the average annual dose to members of the public resulting from all the activities of the nuclear power industry is now less than one millionth of a sievert (1 $\mu$Sv) a year (NRPB 1988), about one two-thousandth of the average annual dose from the natural background. On the basis of current risk estimates for low doses and dose rates about two out of the total of about 140 000 cancer deaths a year could be ascribed to these activities.

Although the overall effects of the routine operations of the nuclear industry are thus extremely small, there is a need to look carefully at areas around nuclear facilities where higher-than-average radiation levels might exist and where any associated health hazards may be expected to show up. The absence of any clear indications of excess cancer rates among workers in the nuclear industry, where doses are in general above those received even by the most exposed groups of the public, make it unlikely that any clear pattern of cancer excesses will be found and this indeed is the conclusion of a major study carried out in the UK (Cook-Mozaffari et al. 1987) This very detailed study has been summarized by Forman et al. (1987), who concluded that 'there has been no general increase in cancer mortality near nuclear installations in England and Wales during the period 1959–1980'. The study showed that areas around nuclear sites had lower cancer mortality rates than control areas more often than the reverse. 'This remains true if attention is concentrated on those types of cancer that have been particularly associated with ionizing radiation, namely leukaemia, bone cancer and multiple myeloma.'

The situation is not so clear-cut when it comes to effects in children, where the studies of radiation workers are less relevant. Concern has centred on the unusual clusters of childhood leukaemias that have been detected near the Sellafield and Dounreay reprocessing plant. On the basis of current assessments, radiation exposures in these areas have been far too low to account for the apparent excesses of leukaemia (COMARE 1986, COMARE 1988). Kinlen (1988) has postulated that the childhood leukaemias may have an infective cause, made more likely by a sudden influx of newcomers into a previously well protected community. Such population mixing has certainly occurred as a result of the initial isolation and extreme population influx into the Sellafield and Dounreay areas, and Kinlen found a significant increase of childhood leukaemias following the building of the new town of Glenrothes in a relatively isolated local authority area of Scotland. More recently Kinlen has extended his study to all post-war new towns in Britain. Nine of these were built in areas around London or Glasgow and five (including Glenrothes) were built in largely rural areas further away from conurbations. Significantly increased rates of leukaemias in the under-fives were found in the five isolated towns whereas there was no increase in the towns near conurbations. This finding clearly supports Kinlen's hypothesis and he has concluded that 'strong reasons would be required for supposing that this effect did not operate near nuclear reprocessing sites, so unusual is their demographic pattern.' (Kinlen 1989).

It is interesting to note that ICRP appears to take a far more restrictive attitude to radiation exposure resulting from industrial activities, such as nuclear power generation, than to avoidable radiation exposure from natural sources. ICRP (1977) states that its recommended limits are not intended to apply to 'normal' levels of natural radiation, but only to those components of natural radiation that result from man-made activities or in special environments. Living in a house which is made of building materials containing naturally radioactive nuclides or which has a high level of radon gas because of restricted ventilation are given as examples of activities that can lead to an increase in 'normal' exposure to natural radiation. Yet in its latest recommendations relating to radon in houses ICRP recommends that no action needs be taken unless annual doses from radon in houses exceeds 20 mSv, a factor 20 times greater than the general dose limit for members of the public (ICRP 1984). It is recommended that the annual doses from radon in new houses should not exceed 5 mSv.

In Britain there are over 40 000 people living in houses where the annual dose from radon is over 20 mSv a year, and about 400 000 in houses with over 5 mSv a year (NRPB 1987c). Only about 900 workers in the nuclear industry receive annual doses of over 15 mSv a year with about 4000 receiving between 5 and 15 mSv a year. Even those members of the public most exposed to radiation from the nuclear industry get doses that are over ten times lower than the 5 mSv a year level. The total number of people in Britain receiving doses more than 0.1 mSv above background from the activities of the nuclear industry is probably only a few dozen.

## The nuclear fuel cycle

The use of nuclear fuels in reactors is part of a sequence of operations, all of which have some potential environmental impact. It is necessary to obtain and refine the raw fuel material, and to fabricate the fuel elements before using them in a reactor. As the fissile isotopes are used up by the fission process, the spent fuel has to be discharged from the reactor. The major source of radioactivity in the whole fuel cycle is the fission process itself. The fission products—the lighter atoms produced by the splitting of the uranium nucleus—are radioactive and the level of radioactivity of spent fuel is so high that the fuel elements will be generating considerable amounts of heat even after discharge from the reactor. Spent fuel must therefore be stored with artificial cooling, usually under water. Eventually, however, the spent fuel must be reprocessed to recover useful fissile material (see p. 137) or packaged for final disposal. All these steps, as well as the reactor operations themselves, give rise to radioactive effluents or to radioactive waste. The environmental effects depend on the type of fuel cycle which is used and on the quantities of material to be processed in each stage. The extent of the differences can be illustrated by comparing a fuel cycle typical of thermal reactors with one typical of a fast reactor which is designed as a breeder, as described on p. 80.

Most thermal reactors operating today use an oxide fuel enriched in Uranium-235, and an enrichment stage therefore has to be inserted before the final fabrication of the fuel. A fuel cycle for a pressurised water reactor (PWR) is shown in Fig. 3.1 (Flowers et al. 1986). The quantities shown are the approximate amounts required for the generation of 1 GW of electricity, in metric tonnes. The spent fuel, uranium oxide in a zirconium alloy, can be stored almost indefinitely under water and can then either be reprocessed to recover the uranium and plutonium for future use, or consigned directly to permanent disposal after suitable packaging (see p. 155).

The series of magnox reactors which are operating in this country use natural uranium metal fuel, and therefore no isotopic enrichment stage is required. Uranium metal is highly reactive, as is the magnesium alloy in

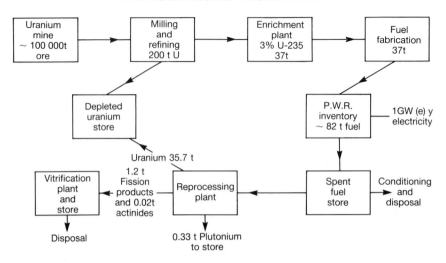

**Fig. 3.1** The flow of fuel material through a typical pressurized water reactor generating 1 GW of electricity for one year.

which it is canned, which gave the magnox reactors their name. The spent fuel is therefore not suitable for prolonged storage underwater or for permanent disposal. Reprocessing has always been seen as an essential part of the magnox fuel cycle and seems likely to continue for the remainder of the lifetime of the magnox reactors, which may be a further ten to fifteen years. The only other commercial reactor which uses natural uranium fuel is the Canadian CANDU type, but the fuel is uranium oxide canned in a zirconium alloy, which is chemically more stable and therefore more easily stored; no decision has been taken yet whether or not to reprocess the spent Canadian fuel.

Plutonium can be used in place of U-235 in thermal reactor fuel cycles and much work to that end is in hand. Typically, the requirement for fresh uranium can be reduced by 15–25 per cent by recycling the uranium and plutonium. However, it is more efficient to use the plutonium in a fast reactor cycle, which is designed to 'breed' plutonium from U-238. The amount of new uranium required to keep the reactor in operation is about one-fiftieth of that required in a thermal reactor fuel cycle, and the depleted U-238 already in store in this country is a source of energy roughly equivalent in size to our coal resources. A fuel cycle typical of a large plutonium-fuelled, sodium-cooled fast reactor is shown in Fig. 3.2. The reactor core is surrounded by a 'blanket' of U-238, to 'breed' more plutonium. A reprocessing step is essential to the economy of this type of fuel cycle, because of the need to recover the large amount of plutonium from the spent fuel (Flowers *et al.* 1986).

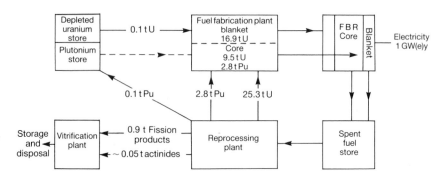

**Fig. 3.2**   The flow of fuel material in a mature fast reactor fuel cycle, based on generating 1 GW of electricity for one year from a liquid-metal-cooled fast breeder reactor (LMFBR).

On the national scale, the contribution from all stages of the fuel cycle to the radiation dose suffered by the population at large is small, amounting to less than 0.1 per cent of natural background (Hughes 1988). On the global scale, the impact is also small. UNSCEAR collects results on exposures to ionizing radiation from all sources; the total collective dose commitment due to the world's nuclear programme up to 1982 amounted to one day of exposure to the average natural background of radiation. Radiation doses occurring to people living close to nuclear plants were naturally higher than the average figure and details are given in the succeeding paragraphs as individual parts of the fuel cycle are discussed.

The regulation of emissions from all nuclear plants in this country is the responsibility of the Environmental Departments and of the Ministry of Agriculture, Fisheries, and Food, as has already been explained (p. 100). The safety of operating plant is primarily the legal responsibility of the plant operator. The public interest is protected by two independent bodies, the Health and Safety Commission and the Health and Safety Executive (HSE). It is the duty of the HSE to develop the principles to be followed in assessing and reducing risks and to monitor operations to ensure compliance.

Every commercial nuclear station has to be licensed by the Nuclear Installations Inspectorate (NII) of the HSE, a body which can monitor compliance with safety regulations and refuse to maintain or extend an operating licence if standards are not satisfactory. Apart from setting numerical safety standards, the primary principle enunciated under the relevant Acts is that an employer should do whatever is reasonably practicable to reduce risk—the As Low As Reasonably Practicable principle (ALARP), which is the analogue of the ALARA rule established in radiological protection. Particular examples of the work of the NII are discussed in succeeding sections. Somewhat similar independent bodies exist in other countries with the power to license nuclear plants—e.g. the Nuclear Regulatory Commission in the USA.

## Mining and refining uranium

Uranium is widely distributed in the world's crust with an average concentration of 203 parts per million. It is concentrated well above normal levels in some rocks such as granites (0.001 to 0.002% uranium) and phosphate rocks (0.01 to 0.02%) but economic recovery is seldom possible below 0.05 per cent. Uranium reserves at production costs less than \$80/Kg U were estimated by the IAEA at 1.7 million tonnes in 1985, and raising production costs to \$130/Kg would add another 0.6 million tonnes. In the Western World, the largest reserves occur in the USA, Australia, South Africa and Canada, with other important quantities in Namibia, and Brazil (Spaargaren 1988). Some uranium has been mined in Cornwall, but none is mined at present in the UK. A rough comparison of useful ores with total quantities in the earth's crust is given in Table 3.4. The annual production now is about 40 000 tonnes a year, with a projected rise to 100 000 tonnes in the year 2000. Since a typical uranium content is 0.2 per cent, the production of 40 000 tonnes of uranium involves mining 20 million tonnes of ore. Most of the early mining consisted of opencast technology but conventional underground mining is also employed at depths between 300 and 3000 metres.

**Table 3.4**
Approximate quantities of natural uranium (Roberts 1984)

| | |
|---|---|
| Uranium in top 1 km of earth's crust: | 4. $10^{12}$ tonnes |
| Uranium in the world's oceans | : 4. $10^9$ tonnes |
| Uranium in ores | : 5. $10^6$ tonnes |

The uranium is recovered from the ore by first crushing and grinding to a small particle size, and dissolving out the uranium using either an acid or alkaline leach followed by chemical purification. The object is to leave behind all the undesired constituents of the ore in the milling tailings. The physical characteristics of these tailings is determined largely by the grinding process, which reduces the ore to particle sizes below 0.5 mm, and often much finer. The leached liquor is subjected to purification processes such as ion exchange or solvent extraction to remove the uranium, often in a kerosene solution, and the uranium is finally washed back again into an aqueous phase and precipitated as ammonium di-uranate ('yellowcake') by adding an alkali.

The choice of an acid or alkaline leach makes relatively little difference to the radiological impact on the environment, although the impact of the operations at the mill will differ according to the quantities and types of chemicals employed. The radiological effects are due mainly to the radioactive members of the decay series that starts with U-238. Smaller contributions occur from the decay series of the lighter isotope, U-235, and from any thorium that may be associated with the uranium ore. The U-238 series is summarized in Fig. 3.3; after 13 separate steps, the final, inactive product is an isotope of lead, Pb-206. During geological time, the rate of radioactive decay of all members of the series, in disintegrations/sec, will have become equal; the radioactivity of the original ore body will therefore be about 13 times that of the U-238 content. With the removal of most of the uranium, and the decay of short-lived daughters, the residual activity in the mill tailings will be some 70 per cent of that in the original ore. This will decay with the half-life of Th-230, which is $2.5 \times 10^5$y. After a period of about a million years, the activity will be dominated again by that of the daughter products of the uranium left behind in the mill tailings, perhaps 1–10 per cent of the original uranium in the ore.

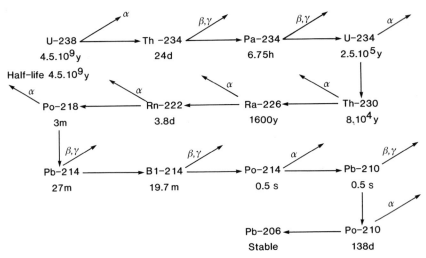

**Fig. 3.3** The radioactive series starting from the most abundant isotope of uranium, U-238.

The mill tailings therefore represent a considerable source of long-lived radioactivity. Although it will all have been present in the ore body, the processes of mining and grinding will have resulted in a product which is more easily dispersed and which therefore constitutes a greater potential hazard. This was not realized in the early days of uranium mining and tailings were used in some cases as backfill and even as construction materials. However, modern practice is aimed at achieving high standards of health and environmental practice during milling operations, and at the satisfactory containment of the tailings for a long period after a mine has been worked out, to reduce any resulting radiation dose in accordance with the ICRP principles for as far into the future as is practicable. The most important nuclides are radon, Rn-222, a gas which can escape from the tailings and be dispersed, causing irradiation of the lungs by its short-lived, airborne alpha-emitting daughters (see p. 101), and the more soluble species such as radium, Ra-226, and lead, Pb-210.

Radiation doses to the public are dominated by those arising from the mill tailings, except for those living in town sites close to mining operations. They can arise from the emission of radon, from the transport of dust due to erosion and from soluble nuclides transported in water. The modern treatment of mill tailings consists in using natural features of the landscape and additional engineering construction to provide as complete containment as possible; the optimum choice will depend on the local climate and on the terrain. In arid areas, it may be possible to store the tailings above the groundwater level; in high rainfall areas, tailings can be stored below groundwater level if the walls of the containment are impermeable or treated with lining materials to reduce seepage.

The most usual technique is to deposit the tailings behind an engineered dam wall, using some natural features of the landscape to form part of the containment if possible. The tailings are then pumped into this area and allowed to dry out in place. Again, steps have to be taken to reduce seepage or overflow to natural water and to reduce the rate of emission of radon from the surface of the tailings, either by maintaining water cover or adding a few metres of soil and rocks on top of the tailings when they have dried out and compacted (IAEA 1981).


## Radiological hazards

Uranium mill tailings can be regarded as a special case of a large volume of waste of low activity per unit volume, which contains in total significant quantities of long-lived radioactive nuclides. The aim of the containment technology and siting adopted is to leave the tailings as a stable, consolidated mass so that environmental protection is provided by passive means and no further active intervention is required. Estimates of radiation doses from the tailings in the future are made after due measurements have been made on and around the site. Such calculations depend upon an estimation of dose from every pathway to an actual or projected population living within a reasonable distance from the site. The emission of radon gas and of liquid effluent must be considered. The emission of radon can cause irradiation through the following pathways:

Atmospheric
(1) inhalation of radon and daughters;
(2) inhalation of airborne particulates;
(3) external radiation.

Terrestrial deposition
(1) ingestion of contaminated foodstuffs;
(2) external irradiation.

The slow leaching of soluble radionuclides from the tailings can cause further irradiation through aquatic pathways, namely:

(1) ingestion of contaminated water;
(2) ingestion of irrigated foodstuffs or of fish;
(3) external irradiation.

The gamma radiation dose at 1 metre above an uncovered pile of typical mill tailings, from an ore containing 0.2% or uranium oxide, $U_3O_8$, would be of the order of 0.01 mSv/hour, so that anyone staying in such a position would receive an additional dose equal to the annual natural background exposure in about eight days. So access to a mill tailing site should be restricted, and the tailings should not be used as backfill or as construction material.

The direct gamma radiation dose will fall off very quickly with distance from the source. However, doses due to radon are more widely distributed since radon, a gas, can be transported over large distances. The increase in levels due to the release of Rn-222 from uncovered dry tailings at a distance of 2 km from the source has been calculated to be of the same order as the natural background resulting from the radium concentration in ordinary soils (IAEA 1981). In practice, a cover of a few metres of clay and topsoil on top of the tailings pile should reduce the exhalation of radon by a factor of five to ten, and the doses due to it would then be indistinguishable from the variation in natural background from place to place.

The calculations of estimated dose to the surrounding population have to be carried out for each site, and repeated for different measures that might be taken to reduce specific pathways. The costs of implementing each option can then be compared with the estimated beneficial effect and a course of action that satisfies dose limitations and the ALARA principle can be determined by discussion with the regulatory agency in the country concerned. On a short time-scale, the magnitude of individual doses to the most exposed individual or to a critical group may be the most significant factor affecting decisions on tailings containment, though the numbers of people affected should be low, since most uranium mining occurs in regions of low population (Uranium Institute 1984).

UNSCEAR has carried out calculations of the collective effective dose and dose commitment arising from uranium mining and milling. A model of a mill environment was assumed, for a mill processing 600 000 tonnes per year of 0.2 per cent grade ore, producing enough uranium to produce 5 GW of electricity a year for twenty years. A population density of 25 per km$^2$ was assumed, out to 2000 km from the mill, a figure which was arrived at after considering the actual population densities in mining areas. The total collective dose commitment to the local and regional population was estimated to be 54 man-Sv for the twenty-year live of the model mine. For comparison, the collective dose to that assumed population from natural background in twenty years would be 600 000 man Sv.

The pathway analysis cannot be carried out with any precision for times greater than some thousands of years, because of increasing uncertainty concerning climatic and geological changes, which can affect the rates of erosion of cover materials or the permeability of surrounding strata. Radon emissions will continue due to the residual uranium content, and global circulation will mean that large populations could receive minute radiation doses from emissions from mine tailings. Estimates of the collective dose are extremely uncertain over long time-scales because of unknown changes in climates, in population distribution and in habits. Further, after thousands or tens of thousands of years, the tailings may well become dispersed owing to the combined influences of climate and geological movement. There would then be some enhanced dose commitment due to dispersion in the aquatic or marine environment. UNSCEAR has calculated that the total dose commitment for releases over 1000 years would be a minor addition to background.

To summarize, techniques are now available which can be applied to reduce radiation doses arising from uranium mining and mill tailings to below the levels recommended by the ICRP so long as the surface geology remains stable. The collective dose that could accrue in the future from well-managed tailings to the regional or world population will be low. But individuals occupying an old tailings area or using this material might be exposed to higher doses and some thought should be given to discouraging these practices for as long as possible. Precise estimates of individual doses in the very distant future cannot be made, but any additional collective dose will be a small addition to background levels. The treatment of uranium mill tailings and the potential hazards arising from them will be compared with those of other categories of radioactive waste at the end of this chapter.

## Mining hazards

The people most exposed to radiological hazard as a result of uranium mining are the miners themselves. An occupational hazard has been an increased incidence of lung cancer, which has been attributed to the radiological hazard arising in these mines. This may be due to the increased levels of gamma-radiation, to the inhalation of radon and radon daughters, or to the inhalation of radioactive dusts. Up to 30 per cent of miners in underground uranium mines in the US have been reported as receiving 20 mSv a year, while annual radon doses in Canadian mines have fallen from about 300 mSv in 1956 to less than 2 mSv in 1980.* It is worth noting that the most exposed group of all UK workers in 1984 were miners in metalliferous mines, who received an average dose of 26 mSv; such mines are often associated with enhanced levels of uranium (Hughes and Roberts 1984).

Bromley (1986) has made a direct comparison of the hazards of mining coal and uranium. Deaths due to accidents in uranium mines and the risk due to radiation vary very considerably according to the type and location of a mine. In most cases, the risks of accidents are far larger than the estimated risks due to radiation dose. The accident rates suffered in providing uranium are smaller than those in providing coal for a given generation of electricity because of the much smaller quantity required— roughly 100 000 t of ore per Gw(e) instead of 3 000 000 of coal. Counting all risks, Bromley considered that the fatalities incurred in mining coal in different countries were between ten and thirty times greater than those due to mining uranium for the same output of electricity.

## Enrichment and fuel preparation

The principles of the methods used for enrichment of the uranium fuel in the fissile isotope, U-235, have already been described. In this country, the plant at Capenhurst now uses the centrifuge technology and the older diffusion plant is being dismantled; in either case, the feed material used is uranium hexafluoride, $UF_6$ which is transported as a solid and which is easily volatilized as a gas. The hexafluoride is manufactured from uranium ore concentrate at Springfields, where fuel fabrication is also carried out.

At Springfields the ore concentrate is first purified and then converted to $UF_6$ by a series of chemical reactions involving the use of hydrogen, hydrofluoric acid and fluorine. The $UF_6$ returned from Capenhurst after enrichment is converted back to uranium dioxide, $UO_2$, by reaction with steam and hydrogen, and the dioxide power granulated and made into pellets of oxide fuel by pressing and sintering at high temperatures. The oxide pellets are then used to fill tubes of stainless steel to become fuel pins for AGRs, or tubes of zircalloy to make PWR fuel elements. The details of pellet and fuel pin manufacture are carefully optimized to achieve good

* The radiation dose in mines is usually reported in units of the 'Working Level Month', or WLM. The conversion factor assumed here is that one WLM of dose due to the radon daughters is equivalent to 10 millisieverts (mSv).

performance in operation. Fuel for the magnox reactors is uranium metal, made by reducing the intermediate, uranium tetrafluoride, with magnesium metal at high temperatures. The uranium metal is cast into moulds, machined to size and fed into the magnox fuel element cans.

A fuel manufacturing plant is similar to a conventional chemical factory handling acids and gases, and the usual safety rules apply. The radiological hazard arises solely from uranium compounds, with some remaining fraction of the radioactive daughters in the uranium decay series which remain in the ore concentrate. The limiting safety factor usually depends on the chemical toxicity rather than radiotoxicity of uranium. Springfields discharges liquid effluent into the tidal waters of the River Ribble, and boreholes in the site are monitored as well. Radiation doses to critical groups in the area, and also near the Capenhurst Works, are very low, and mainly attributed to discharges spreading from Sellafield.

UNSCEAR (1982) recorded similarly slow effluents from enrichment and fuel fabrication plants elsewhere, mainly in the USA and in Sweden, and used them to determine the average radiation doses from a typical 'model' plant processing 10 000 t of uranium a year. The collective absorbed dose commitment is very small and much lower than the dose commitment due to mining and milling uranium ore, by a factor of at least 10 000.

The fabrication of fast reactor fuel from plutonium and uranium oxides is undertaken at Sellafield. This is an operation that has to be carried out in sealed facilities, because of the hazard of inhalation of plutonium, and emissions from the plant contain plutonium and are filtered before discharge. They are considered together with all other emissions from Sellafield under 'Reprocessing' on p. 139.

## Nuclear reactor operations

The operating 'core' of a thermal reactor is the array of fuel assemblies arranged inside the material chosen to moderate the neutron energies (p. 78). It is quite a small structure; for example, the core of the PWR being built at Sizewell consists of 193 fuel assemblies, each containing 264 fuel rods, in a reactor vessel which is 13.5 m high and 4.39 m inside diameter. Arrangements must be made for the heat to be extracted by a flow of coolant, gas, or liquids, through the core and for control rods of neutron-absorbing material to penetrate the core to control the neutron population. A 'fast' reactor is even more compact; there is no moderator and the core consists only of fuel assemblies and coolant. The core of a large, 1000 Mw(e), fast reactor would be a cylinder about 1 m high and 2 m in diameter.

Nuclear reactors operate on essentially closed cycles: there is no massive emission of gaseous wastes as there is from burning fossil fuels. But all nuclear reactors give rise to some gaseous emissions and to liquid effluent of low activity. The details depend on the type of reactor involved; we shall concentrate on those of prime commercial interest in this country, and mention others for completeness.

The magnox and Advanced Gas Cooled (AGR) reactors have graphite moderators and are cooled with carbon dioxide gas under pressure. The magnox reactors built between 1962 and 1966 in the UK have steel pressure vessels, while the later ones have pre-stressed concrete pressure vessels. The uranium metal fuel has to be run at low temperatures; the gas outlet temperature is between 340 and 410°C. In the later designs, the reactor core, gas circulating circuit, and steam boilers are all enclosed in one pre-stressed concrete vessel. Fuel element failure is rare, about one per reactor per year. The AGRs use enriched oxide fuel clad in stainless steel; all have pre-stressed concrete pressure vessels, and higher gas temperatures and pressures, typically 42 bar pressure with an outlet temperature of 645°C. AGRs have the steam boilers inside the concrete pressure vessel, of 5 m thickness, which has a gas-tight steel liner. There is an external gas treatment plant to maintain the correct composition of the circulating gas.

Some airborne activity is discharged from these stations due to the neutron activation of impurities and to the loss of carbon dioxide containing some carbon-14; the annual collective dose from all the UK nuclear power stations has been assessed as 5 man-Sv, mainly due to C-14 (Hughes 1988). This is 0.004 per cent of the annual collective dose from background radiation. The maximum annual dose to any member of the public arising from AGR stations has been estimated as less than 0.1 mSv (Dale 1982).

Radioactive liquid discharges arise mainly from the storage ponds of spent fuel; since most of the stations are on the coast, the resulting doses to critical groups are low and lower than those attributed to the dispersion of effluent from Sellafield. An exception is the station at Trawsfynedd, where liquid effluent is discharged into a lake. A dose to a critical group of people eating fish caught in this lake was estimated as 0.35 mSv in 1982, reducing to 0.25 mSv in 1987 (Hunt 1988).

Most of the world's reactors use ordinary light water under pressure as both the moderator and the cooling medium. In the most common type—the pressurised water reactor (PWR)—the core is cooled by water at pressures of about 150 bar, with outlet temperatures around 300°C, and contained within a strong steel pressure vessel, some 20 cm thick. The high-pressure water circulates through a steam generator, and steam from this generator circulates through a secondary circuit to the turbines. The reactor core, primary circuit, and steam generator are surrounded by a thick concrete containment. In an alternative concept, the boiling water reactor (BWR), the primary circuit is operated at a pressure and temperature which allow limited boiling to occur within the reactor core, typically 70 bar at 290°C, and the steam passes directly to the turbine outside the containment vessel. It is a simpler arrangement, but carries the penalty that radioactivity may be carried by the steam to the turbines.

PWRs and BWRs are refuelled annually, when about one-third of the fuel is changed. Some fission product activity may leak into the water circuit in the unlikely event of a crack in a fuel element and there is also some transfer around the circuit of radioactive species, such as cobalt-60 formed by the neutron irradiation of steels, because of the corrosive properties of water at high pressure. A fraction of the cooling water therefore has to be diverted through a purification system. Gaseous effluents are eventually discharged from this purification system, and consist mostly of krypton and xenon which have little biological significance. Bush (1982) has analysed the data for 1978 relating to PWRs; dose rates at the site boundary were less than 0.05 mSv a year, while effluent releases might cause a dose commitment of 0.005 mSv a year to the most exposed adult individual and about 0.015 mSv to the thyroid of infants. For comparison, the maximum individual radiation dose to a member of the public resulting from the operation of a PWR at Sizewell was estimated to be 1 per cent of background levels, which amounts to 0.02 mSv a year (Layfield 1987).

The Canadian CANDU reactor uses heavy water as the moderator, with the fuel within zirconium pressure tubes penetrating a tank of heavy water, and heavy water pumped through the pressure tubes to cool the fuel. Tritium is formed in the heavy water, some of which leaks from the reactor core. These reactors characteristically emit comparatively high levels of tritium both as gas and in liquid effluent.

The data that are available from the operating experience of a large number of reactors in different countries enable us to make reasonable predictions of the radiological impact of reactor operation. The surveys made by UNSCEAR and, in the UK, by the NRPB, show that the individual and collective doses from the normal operation of all types of commercial reactors will be low; local actions can certainly be taken to reduce any emission thought to be hazardous. But it has been recognized from the very early days that reactors contain a very large inventory of radioactive materials and that any accident which resulted in an appreciable fraction being released would have serious consequences. So a large international effort has been devoted to the design of reactors and of safety systems aimed at reducing the likelihood of such releases as far as possible.

## Reactor safety

More than 99 per cent of the radioactivity in the core of an operating nuclear reactor is locked up in the fuel elements themselves. Much is actually within the crystalline structure of the fuel material, for example the uranium oxide; the fission product gases and the more volatile elements segregate in the gaps between the fuel and the container, or 'can'. The fuel cans will be in a structure which is part of the primary circuit of the reactor, which is surrounded by a pressure-vessel to contain the cooling fluid. The pressure vessel is surrounded by a thick concrete biological shield; water-cooled reactors are surrounded by containment vessels designed to retain the contents of the pressure vessel if it should break. All these barriers have to be broken or bypassed before a major escape of radioactive material can occur. This could happen in one of two ways:

(1) severe damage to the entire structure caused by an external event such as an aircraft crash or a major earthquake;
(2) the fuel overheating to such an extent that the fuel elements are damaged, releasing radioactivity to the primary circuit, and the circuit itself and the containment being damaged at the same time.

External hazards have to be assessed in terms of their severity and likelihood of occurrence, and the structure then shown to be adequately protected against them. The recent case for building a PWR at Sizewell included a consideration of earthquakes, extreme wind loading, extreme tides and tidal surges, rainwater flooding, extreme ambient temperatures, aircraft crash, and external industrial hazards such as shipping accidents near the coast. For example, the reactor was required to be stable against an earthquake of a severity that might be expected to occur in that region once in 10 000 years, and the probability of an aircraft crash damaging a vital part of the system was considered by looking at the statistics of aircraft accidents and at the ability of the very solid concrete structure to withstand them.

Internal hazards that could lead to fuel overheating can arise from a sudden increase in power or from loss of cooling. The power level of a reactor depends on the neutron population, which is controlled by the control rods. These consist of neutron-absorbing material such as boron or cadmium alloys, and are inserted into the reactor core between the fuel elements; control rods are an essential feature of every reactor design. Should the ordinary control rods fail to act, the control of the neutron level and hence of the power level is maintained by other means of inserting neutron-absorbing material into the core in an emergency; the lack of such provision was one of the contributory causes of the reactor accident at Chernobyl, which is discussed below (p. 129). The second major requirement is to maintain the ability to cool the fuel even after the nuclear reactor has been shut down, by inserting the control rods or other neutron absorbing material, because of the continuing generation of heat caused by the radioactive decay of the fission products. Immediately after shutting the reactor down, this decay heat is about 7 per cent of the total heat output of the reactor—a considerable amount, over 200 Mw from a reactor delivering 1000 Mw of electrical energy. Owing to the rapid decay of the short-lived fission products, the decay heat reduces to 1 per cent of the original heat output after an hour and to 0.1 per cent after six months. Every reactor therefore has to be equipped with an emergency core-cooling system capable of maintaining the integrity of the fuel elements by keeping the temperature within safe limits for a considerable time.

### Safety criteria

Criteria against which plant should be designed, and eventually licensed for operation, must set standards of safety that are acceptable in comparison with other industrial activities. Accidents will happen; instruments and plant components will fail and wear out, and operators, being human, will occasionally make mistakes, particularly if they are under stress. Reactors must be designed so that the consequences of such failures are tolerable which, in practice, means that the probability of an accident with serious consequences—injury, loss of life, or social disruption—must be reduced to very low limits. The word 'risk' is defined in safety assessments to mean the probability of an accident leading to certain consequences in terms of deaths or serious injuries, and a distinction is drawn between risk to an individual and 'societal' risk which is often interpreted as the number of people possibly affected by a single event multiplied by the probability of its occurrence, though other measures of social disruption must be taken into account.

The philosophy of radiation protection has been explained earlier in this chapter and those rules govern the normal operation of reactors. Indeed, more stringent local targets are often adopted; the CEGB design target at Sizewell was that the predicted annual dose to members of the public resulting from Sizewell 'A' and 'B' stations should be assessed against one-thirtieth of the ICRP-recommended limit of 5 mSv—i.e. 0.17 mSv a year.

The definition of safety criteria applicable to accidents has been a matter of vigorous debate for a decade or more. A range of values that is often quoted is that individual risks of death greater than one in ten thousand a year are unacceptable while risks below one in a million a year are likely to be tolerated. The Inspector's conclusion at the end of a very long debate on safety criteria at the Sizewell Inquiry was that a level of individual risk of death of the order of one in a million years is 'likely to be broadly tolerable if justified by associated benefits'; he recommended that further work should be done on specifying social risk criteria and that these should be subject to public and political consideration.

Clearly there are difficulties in defining the risk to society due to radioactive releases when some of the consequences would be long-term and would be the result of the exposure of large numbers of people to small individual risks. The criteria set by the CEGB as design targets deal not with risk but with a maximum frequency of accidents. These targets are:

(1) the frequency of any single accident that could give rise to a large, uncontrolled release of radioactivity should be less than one in ten million a year;
(2) the total frequency of all accidents that could lead to uncontrolled releases should be less than one in one million a year;
(3) the predicted frequency of accidents from which radiation doses of 1 ERL (Emergency Reference Level) could be expected should be less than one in ten thousand a year.

The Emergency Reference Level is that level of radiation dose at which actions for the protection of the public, such as evacuation, have to be considered. The lowest level is a dose of 100 mSv, twice the maximum current annual dose for radiation workers and corresponding to an additional risk of fatal cancer in later life of about three in a thousand. The risk to any individual of such a release can be calculated (as frequency × consequence) to be three in ten million per year, if condition (3) above is satisfied.

Following the recommendations of the Inspector at the Sizewell Inquiry, the Health and Safety Executive has published a paper on *The toleraility of risk from nuclear power stations* (HSE 1987) in order to form a basis on which public, expert, and Parliamentary opinion could be expressed. In this document, the HSE explains the general philosophy behind its application of the ALARP principle to the regulation of risk. This is applied between a level of risk seen as intolerable and a lower level of risk deemed to be broadly acceptable. The control exerted by the Nuclear Installations Inspectorate (NII) is applied prior to and during design, during construction, during testing and commissioning and during operation and eventual decommissioning. It remains the responsibility of the licensee to ensure that risks are reduced according to the ALARP principle.

The HSE suggests that the maximum levels of risk that should be tolerated for any individual member of the public from any large-scale industrial hazard should be 1 in 10 000 per year, while a lower, broadly acceptable level might be one in a million per year. Individual risks arising from nuclear plant that conform to the NII's Safety Assessment Principles should be lower than that. Societal risk is discussed in terms of the tolerable frequency of accidents that affect substantial numbers of people. The HSE notes that the design specification of the Thames Barrier was that the chance of its being overtopped by a freak tide should be less than 1 in 1000 a year, which is also the predicted approximate annual chance of an aircraft crash killing 500 or more people in the UK. After an earlier inquiry into the risks potentially arising from the petrochemical complex at Canvey Island, the HSE had calculated a chance of about 1 in 5000 per annum of a major accident. On the basis of these comparisons, the HSE considers that the frequency that might be accepted as tolerable for a serious nuclear accident anywhere in the UK might be 1 in 10 000 per annum. A serious accident in this context is defined as one that might give rise to an estimated 100 later deaths from cancer; larger accidents should be less frequent. This standard is compatible with a goal of one in a million a year for an uncontrolled release of radioactivity from a single reactor. The HSE points out that this order of societal risk makes only a marginal addition to the risks arising from other activities.

Broadly similar safety targets have been adopted by other European countries (J.C. Consultancy 1986). In the US, the Nuclear Regulatory Commission had adopted two qualitative safety goals:

1. Individual members of the public should bear no significant additional risk to life and health.
2. Societal risks to life and health from nuclear power plant operation should be comparable to or less than the risks of generating electricity by viable competing technologies and should not be a significant addition to other societal risks.

The Commission adds the quantitative statement that the risk of early death should not exceed 0.1 per cent of the average accident rate and that the risk of latent cancer mortality should not exceed 0.1 per cent of the cancer mortality risk from other causes. The cancer risk is to be averaged over the population within a 10-mile radius of the plant. Since the average risk of dying from cancer is about 3 in 1000 per year, the maximum additional fatal cancer risk should be three in one million per year. They further state that the mean frequency of a large release of radioactive materials from a reactor accident should be less than one in one million per year of reactor operation, which is similar to the target adopted by the CEGB (Okrent 1987), and suggested as tolerable by the HSE.

### Principles of safe design and operation

To meet the design targets and safety goals discussed above, reactors have to be designed on sound engineering principles so that each component is either able to give reliable service under the operating conditions of temperature, pressure, and irradiation over the lifetime of the plant or be capable of easy maintenance or replacement. While safety is attained as far as possible by inherent features of the design, reactors must also incorporate safety systems to ensure control of the neutron levels and adequate cooling of the fuel, even under fault conditions. These systems consist of instruments able to detect when some plant parameter has gone beyond safe limits and also automatic emergency shut-down and cooling systems.

The principles on which both reactor instrumentation and safety systems are designed owe much to the concepts of high reliability engineering developed in the aircraft industry. They may be summarized as follows:

1. Redundancy:  There must be adequate protection against the failure of individual components so that at least one safety system is available against demand at any time.
2. Diversity:  Components of a given type may be liable to a common mode of failure; it is safer to include more than one type.
3. Segregation:  Safety systems should be spaced far enough apart so as not to be subject to common cause failure, such as a local fire, and plant layout should minimize the chance of accident propagation.
4. Containment:  Means should be included to confine accidents as far as possible.

The manner in which these principles are applied varies with reactor type. Reactor control rods are arranged in more than one circuit, and the required reliability is achieved by duplication of the sensing instruments and of the control mechanisms; the rods drop into the reactor core under gravity if the electrical supply fails. The action of a fraction of the control rods is enough to initiate a reactor shut-down. If all should fail to act, the core of an AGR can be filled with nitrogen, a neutron-absorbing gas, while a PWR can be flooded with a solution of boron compounds to absorb neutrons.

The rate of rise of temperature under fault conditions of gas-cooled reactors, which have a large mass of graphite moderator, is slow. Faults that could result in a loss of cooling capacity include loss of power supply to the gas circulators and a failure of the primary circuit with release of carbon-dioxide. Should forced circulation not be available, calculations show that sufficient cooling of the fuel will be provided by natural convection, with a large safety margin; independent pumps and water-cooling systems maintain a supply of cooled water to the boilers. Supplies of carbon dioxide are maintained on site so that the pressure in the primary circuit can remain above atmospheric pressure, thus restricting air ingress in the case of a leak in the circuit. The total failure probability of all the groups of faults considered was found to be between 1.5 and 5 per million reactor years of operation and the probability of events that would lead to a release of radioactivity to the environment would be less than that (Dale 1982).

The effectiveness of the emergency core-cooling system is of particular significance to the safety of water-moderated reactors, PWRs and BWRs, since the fuel must be kept covered by water at all times. Steam is not a sufficiently good heat transfer medium to prevent the fuel overheating, and at high temperatures, above 950°C, steam can react with the zirconium alloy used as a fuel canning material to form hydrogen gas. A modern PWR is therefore equipped with a number of high capacity emergency core-cooling systems to meet the possibility of a sudden, large loss of cooling accident. The system is arranged to provide both redundancy and diversity. For example, the design for the Sizewell PWR allows for four high-pressure injection systems, two low-pressure injection systems and four accumulators full of borated water which would be injected into the reactor core if the circuit pressure should fall. There are also two systems for heat rejection in case the main boilers are not available and all emergency systems are serviced by an independent set of emergency electrical supplies.

A sodium-cooled fast reactor of the design of the PFR at Dounreay or of the large Superphenix reactors in France consists of a compact core of fuel immersed in a large pool of sodium, which means that the temperature will rise very slowly under most fault conditions. In addition, loops full of liquid metal leading to air-cooled heat exchangers provide ample means of rejecting heat if the main cooling system should fail. Over-heating of the fuel after reactor shut-down can be avoided by natural convection alone without the need for pumps, electricity, or water supplies (Marsham 1985). A reliable shut-down system is required since events can be envisaged which might lead to a sudden increase in power, though it seems that the containment would not be breached even in such an unlikely event (Broadley 1986; Cogne and Justin 1985).

The requirements for secondary containment, as a final barrier against dispersion, will again differ for each reactor system. No secondary containment is provided for AGRs beyond the massive concrete vessel, since faults in the primary circuit lead to a drop in circuit pressure and cooling of the fuel is relatively easily maintained. There are no sudden discontinuities in the heat transfer conditions such as occur with the change from water to steam in water-cooled reactors, and it is enough to provide some auxiliary enclosure around potential leakage points. A separate containment structure is always provided for PWRs and BWRs, designed to withstand the rapid release of all the coolant contained in the primary circuit. This structure is typically a strong concrete shell with a steel liner; emergency cooling by means of water sprays and fans are fitted inside this secondary containment, together with means for recombining any hydrogen released with oxygen to form water. Some designs are now being fitted with filtered vents to avoid over-pressurization of the secondary containment building, but the advantages in terms of increased safety have to be assessed carefully (Edwards 1985). Sodium-cooled fast reactors are low-pressure systems; they are enclosed in a concrete containment building designed to withstand the effects of sodium fire in the event of a severe break in the primary circuit.

There remains one class of safety question which has to be approached from the point of view of fundamental engineering principles and materials science: that is, the proof of the integrity of the main structures surrounding the reactor core. These include the steel pressure vessels of the earlier magnox reactors and the steel pressure vessel of the PWR or BWR, the reinforced concrete pressure vessel of the AGR and the steel structures which define the internal gas flow-paths in that reactor. It is necessary to demonstrate conclusively that such vital structures will not fail during the lifetime of the reactor. As an example, the safety of the reactor pressure vessel was the subject of very searching examination at Sizewell. Essentially, the case rested on the soundness of the design, materials specification, methods of inspection, and pre-commissioning testing. In addition, there were theoretical arguments that the techniques of inspection to be applied in service would be able to detect cracks which might grow and cause failure of the vessel before they became dangerous. An estimated probability of failure of once in four hundred million years was thought to be realistic or even pessimistic. The Inspector accepted that the safety case was justified, in view of the rigorous programme of inspection by two independent authorities that had been agreed by the CEGB.

**Principles of safe operation**

While a reactor is designed as far as possible to be 'fail safe', the importance of competent, reliable operation cannot be overestimated. It is widely recognized that an operating organization should have in place:

1. A sound training scheme for operators of adequate educational standard, with refresher courses and experience on reactor simulators:
2. A strong safety department indpendent of site management and responsible to the highest levels in the organization. The safety department organization would be responsible for
   — overall safety philosophy and criteria;
   — drawing up operating schedules and rules;
   — specification of safety duties;
   — monitoring and inspection;
   — reporting of any unusual occurrences;
   — formulating emergency plans.

The formal responsibility for safe operation must rest on the organization responsible for the reactor site. But a detailed examination of the regulations, recording, and reporting of safety matters is also undertaken by the licensing authorities in a manner reflecting the accepted norms of government regulation in different countries. It is reasonable that the relations between the Nuclear Regulatory commission and the many utility companies in the USA should be more detailed than those between the

Nuclear Installations Inspectorate (NII) and the two large utility companies, the CEGB and the SSEB, in the UK. The NII sees its role as essentially to monitor the licensee's performance as regards safety, and values the flexibility inherent in the relationship (Anthony 1986).

That relationship came under sharp scrutiny at the Sizewell Inquiry. While accepting that each organization was technically strong and competent to carry out its tasks, the Inspector was critical of the organization of the licensing process and of the efficiency of the exchange of information between the NII and the CEGB. Further, the Inspector recommended that the NII should publish more on the way that safety criteria are applied in practice, and that the HSE should take the lead in coordinating the work of the various public agencies engaged in some aspect of nuclear safety (Layfield 1987). The document proposing risk criteria is the first reaction to those recommendations; in several paragraphs, the HSE emphasizes the importance of the 'safety culture' which it expects to find in the operating organization.

It is the licensing authority's duty to ensure that the safety criteria have been met beyond reasonable doubt before a licence to construct and then to operate a reactor can be issued. Methods of assessing the reliability of the safety and emergency systems in a quantitative way therefore have to be developed, and a brief description of these methods follows.

## The assessment of risks

An engineered structure or plant is designed to be safe for its working life by identifying the critical faults that could lead to failure and by ensuring that the structure will not be subject to them, either through design features or by maintenance and replacement. This traditional approach to safe design is known as the 'deterministic' method and it has been and still is used in the assessment of reactor designs. It is also possible to assess the risks that could conceivably arise by a quantitative analysis of the probability and of the consequences of all events that could lead to plant failure. Such an analysis is known as a Probabilistic Risk Assessment (PRA) or Probabilistic Safety Assessment (PSA) and the possible effects of faults initiated both by external events and by internal plant failures can be assessed in this way. A probabilistic approach is also necessary in order to be able to demonstrate that the reliability of the safety systems is sufficient to meet the demands of safety criteria that are expressed in terms of probabilities.

The many possible events that can occur in a complex plant can be broken down in a systematic way into the behaviour of pieces of equipment, and the results of individual operator actions. Probabilistic risk assessment involves, first, the identification of all sources of potential hazards and then the specification of the sequence of events—failures of safety systems or operator mistakes—that could lead to an actual release of radioactivity. Estimation of the risk of a release then involves the calculation of the frequency of occurrence of each event in a sequence that leads to failure. The information on reliability, or of failure to act on demand, of instruments or of plant components is based on past experience and such data are stored in various centres such as the National Centre of Systems Reliability Data Bank, though the information can only be based on a statistical sample and failure rates may not be constant. Judgements on operator action are less reliable, and licensing authorities increasingly expect a sequence of events that could lead to a hazard to be terminated by automatic means rather than by operator action.

The analysis of sequences of events needed in order to perform a PRA is often set out in the form of event trees or fault trees. An event tree traces the possible sequence of events following from one initiating event, while a fault tree traces the various sequences of events that could lead to a particular failure. A simplified fault tree taken from the CEGB evidence to the Sizewell Inquiry is shown in Fig. 3.4; it is the top end of a fault tree illustrating how activity might be released following a loss of cooling

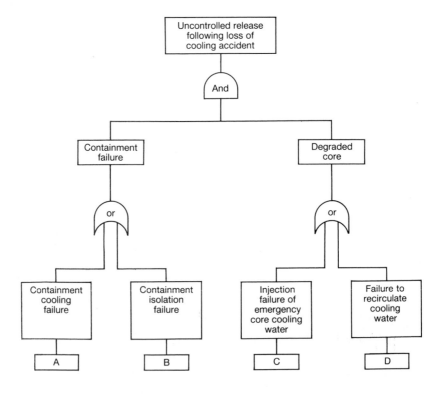

**Fig. 3.4**  A simplified fault tree applicable to a PWR.

accident. This simplified fault tree shows that both the fuel and the containment must be damaged before an uncontrolled release of activity can occur; the containment may be breached because of loss of cooling or failure to isolate, while the fuel in the core may overheat and be damaged if either the emergency core cooling system fails to supply cooling water or the cooling water recirculation system does not work. Further detailed fault trees can then be constructed to illustrate events that could cause failure of any of these four safety systems, and the probabilities of failure combined with the probability of the initiating event—a loss of cooling accident—to obtain the overall probability of producing an uncontrolled release of activity. This type of analysis can be repeated for other initiating faults, and, by adding all the results, the total probability of uncontrolled releases from all causes can be estimated.

One can then carry out a consequence analysis. If the amount of radioactivity released into the atmosphere can be estimated from the circumstances of the fuel failure, the probabilities of subsequent radiation doses to people or of contamination of land can be calculated if the weather conditions and distribution of population are known. Meteorological data can again be expressed in a probabilistic way, and account can be taken of the possibility of taking simple remedial measures such as sheltering indoors or evacuation of people who live downwind of a reactor. Contamination of land can lead to later exposure due to the ingestion of contaminated foodstuffs. By considering all possible pathways, the risk to the population can be expressed as the probability of a given number of people receiving radiation doses of various sizes.

The first complete study of this type was an assessment of accident risks in commercial power stations in the USA made by a team led by Rasmussen (1975). A full PRA was carried out on two reactors, at PWR at Surry, Virginia, and a BWR at Peach Bottom, Pennsylvania, and the results were extrapolated to refer to a nominal programme of 100 large reactors. The study concentrated on accidents involving melting of the fuel ('core melt') since these were the only ones that could affect the public, and took into account both external and internal initiating events. The results were expressed as the frequency of events that could result in more than given numbers of fatalities, and both early deaths due to massive irradiation doses and late cancer deaths due to lesser doses were calculated. These estimates were compared with statistical results for other types of man-made and natural disasters which had been published in 1969.

The Rasmussen Report was published at a time of considerable anxiety about reactor safety in the USA and it attracted considerable criticism. A review group was set up by the Nuclear Regulatory Commission in 1978. This review endorsed the general methods that had been used: the fault tree/event tree approach was seen to provide the most complete single picture of accident probabilities associated with nuclear reactors, and this type of analysis should then be the principal means of dealing with generic safety issues. The main criticisms of the report were that comparisons of the risks of nuclear reactor operation with those arising from very different activities were misleading, in that the risks were of a different type, and that the accuracy of many of the probability factors was not as good as the Rasmussen Report had claimed; the resulting range of values should have been stated more clearly in the summary of the report (Lewis 1978).

Since 1978, more than a dozen PRAs of US reactors have been published, as well as an important study of German reactors, and the use of the technique is spreading to chemical plants. As a result of all this work the uses and limitations of PRA are better understood. The main benefits are the qualitative insights that are gained into the important factors affecting safety in very complex plants; PRA shows how systems interact after failures and the discipline of having to assign quantitative values to specific modes of failure yields valuable guidance on priorities in research and development. But the accuracy of the risk estimates made can only be as good as the accuracy of the information on which they are based and the answers should be expressed as a range of probabilities or be subjected to a sensitivity analysis.

It is clearly impossible to verify the results of complete PRAs by comparison with direct observation or experiment. But the methods used can be cross-checked by comparing the approach made by independent teams, and confidence in the results increases as more exercises are carried out. Further, some parts of the analysis can be tested against experiment. For example, the complex computer codes used to predict what happens as fuel heats rapidly to high temperatures are now based on extensive experimental tests where fuel rods are held under conditions that mimic reactor conditions but that can be varied suddenly. An opportunity to carry out tests on an AGR under extreme conditions arose when the prototype reactor, the Windscale Advanced Gas Cooled Reactor, was to be shut down in 1981 (Bridge *et al.* 1982). These experiments covered the testing of both unirradiated and irradiated fuel pin clusters up to a peak cladding temperature of about 1300°C, which was the peak temperature postulated to occur in some fault conditions. The fuel pins survived in good condition. In addition, the computer codes used in fault studies for large changes in power and flow were tested, and studies made of the release of volatile fission products from failed fuel elements and their deposition in the reactor circuit. The observed behaviour was in general consistent with prior predictions. Tests of this type have also been carried out using fast reactor fuel pins; short review articles have been published by Hennies (1986) and by Broadley (1986).

**Recent risk assessments in the UK**

PRA is now seen as a very valuable qualitative tool, yielding engineering insights concerning safety performance, and as a valuable quantitative method of making risk estimates for those parts of a reactor for which an adequate set of reliability data exists. The use made of PRA in the safety case for the proposed PWR at Sizewell is typical. The CEGB based their case on the identification of a category of initiating faults and fault sequences whose likelihood of occurrence is sufficiently high that their consequences must be evaluated and shown to be acceptable; it must be demonstrated in each case that the reliability of the protection and safety systems is such that the probability of failing to bring the fault sequence to an end is less than 1 in 10 million a year, which is the design criterion. This analysis was carried out with extensive use of event and fault three methods. Faults of even lower probability are 'beyond the design basis'. However, an analysis of the consequences of faults that were so improbable that they lay beyond the design basis was carried out using PRA methods in order to be certain that no sequence with large consequences could be identified just beyond the design basis boundary. These calculations all involved accidents in which the reactor core was damaged; it was estimated that the fuel would melt with a frequency of one in a million years and that accidents involving both molten fuel and containment failure would have a frequency of 7 in 100 million years.

In all, twelve release categories were identified and the radiological consequences for each calculated using meteorological records, and combined to estimate the total risks arising from severe accidents. The individual risk estimates were very low—less than 1 in 100 million/year at 1 km from the reactor, and reducing sharply with distance. The societal risk was calculated using the size of population exposed to these low levels of individual risk. Figure 3.5 illustrates the frequency with which degraded core accidents could occur and give rise to calculated numbers of fatal cancer cases; the frequency of an accident causing 100 cases or more is about 1 in 100 million a year (Ashworth 1987).

The figure of one in a million a year for the frequency of core melting was criticized at the Inquiry as being too optimistic compared to results obtained in other studies. The Inspector's final conclusion was that the annual probability of a degraded core would be unlikely to exceed 1 in· 100 000, but that in most cases the containment would remain intact. The annual probability of an uncontrolled release was therefore unlikely to be more than one in a million. He concluded that the annual social risk, as computed from the likely number of deaths per year, was dominated by the risks from normal operation, since the probability of severe accidents was so low. The figures are collected in Table 3.5. The Inspector's opinion was

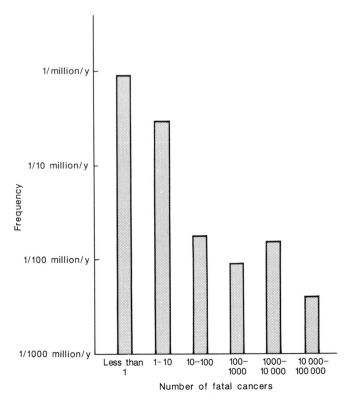

**Fig. 3.5** Estimates of the frequency of 'beyond design basis' accidents at the Sizewell B reactor causing defined numbers of fatal cancer cases in the future.

that this level of social risk 'is a minute risk by any standards for an industrial enterprise of the magnitude and complexity of Sizewell B.'

An important feature of a risk assessment is the illustration of the relative importance of the different release mechanisms. The analysis revealed the importance of maintaining the containment and of a correct estimate of release fractions. As a result, the CEGB has subsequently added a third isolating valve in the relevant pipework to reduce the likelihood of a break bypassing the containment and has increased the systems designed to remove heat from the containment (Ashworth 1987). In addition, much work is still in progress on the estimation of release fractions under practical conditions. Results such as those illustrated in Fig. 3.5 are based on estimated releases which are probably too high, since they underestimate the degree to which fission products could be trapped on material and structures within the containment.

The attainment of high standards of safety is not only a matter of good design and of sound operating rules; maintenance and inspection are also

**Table 3.5**

Estimated social risks to members of the public in the UK from Sizewell B
(Layfield 1987)

|  | Annual social risk (deaths/year × 10 000) |
|---|---|
| Normal operation | 16 |
| Design basis accidents | 2 |
| Beyond design basis accidents | 2 |
|  | — |
| Total | 20 |
|  | — |

That is, on average, 1 death in 500 years

required. Further, improvements in design for safety are continually being made and standards are being tightened. It is necessary to assess older plant against up-to-date criteria from time to time. The Nuclear Installations Inspectorate (NII) has recently published the first of a series of major reviews of the state of the magnox stations, some of which are reaching the end of their design lifetime. The first of the CEGB's stations to be reviewed was Bradwell, which started operation in 1962; the GEGB commenced a long-term safety review in 1982, in order to seek extension of the site's operating licence until 1992.

The overall conclusion of the NII's examination of the case was that the review generally confirms that the existing safety case is valid for the present operation of the reactor. But they have identified several aspects where the CEGB has still to demonstrate that the reactor can be operated to modern standards of safety until 1992—for example, in the standard of in-service inspection applied to welded joints where ducts join the main pressure vessel, and in modifications that could reasonably be made to reduce sources of direct radiation from exposed ducts and pipes. The NII detailed several requirements on which it required more work to be done (HSE 1987). The second case in this series concerned the magnox reactors at Berkeley. These were closed down in March 1989, since the expense of refurbishing them for a longer life was not justified.

### Historical: reactor accidents

There have been two accidents to commercial reactors which resulted in extensive damage to the reactor core and consequent release of activity to the environment, one to a PWR at Three Mile Island, near Harrisburg, Pennsylvania on 28 March 1979 and the other to an RBMK reactor at

Chernobyl in the Ukraine on 25 April 1986. The accident that occurred in Windscale in 1957 was to an early design of military reactor for the production of plutonium, and that type of reactor has since been taken out of service.

The Three Mile Island accident is an example of a small loss of coolant accident which was mismanaged and led to severe damage to the reactor core but to little escape of radioactivity beyond the containment, which was not breached. The original fault was a loss of feedwater which caused a main turbine to trip and the pressure in the primary circuit to increase; the safety systems acted properly to shut down the reactor and a pressure relief valve opened to release the excess pressure. However, unknown to the operators, the relief valve failed to shut when the pressure fell to normal levels, allowing water and steam to escape to the containment building. Over the next 100 minutes, about two-thirds of the water in the primary circuit was discharged to the containment until the operators finally realized that the pressure relief valve was stuck open and closed the leak. But inadequate water flow was maintained to the core and the level of water fell, exposing the fuel, which overheated with subsequent reaction between the zirconium cladding and water, releasing hydrogen. Cooling was finally re-established and the hydrogen slowly vented over the next five days. The radioactivity released to the environment was almost entirely the rare gases, krypton and xenon, and a very little iodine (0.6 TBq). The health consequences will be minimal, up to five possible cases of cancer over the next thirty years, compared with the 540 000 cases that will develop in that population in the same period (Kemeny 1979).

However, considerable mental stress was caused in the neighbourhood, partly because of conflicting technical reports during the early stages and the advice that pregnant women and young children living within five miles of the reactor should leave the area—advice actually given on a mistaken technical evaluation. The accident led to a lowering of public confidence in the nuclear industry and in the NRC. The President's Commission set up to investigate the accident was critical of the communications concerning safety within the utility concerned, of the standard of training of the operators, of the layout of the control room, of some of the operations and regulations of the NRC and of the emergency arrangements. The damage to the reactor was extensive and new techniques have had to be developed to deal with so much active equipment; the clean-up operation is estimated to cost up to $1 billion (Duffy *et al.* 1986).

The accident caused a very radical re-examination of safety-related planning, mainly in the USA but also in other countries. The NRC eventually produced a comprehensive TMI-2 action plan designed to ensure that the lessons would be learnt and applied. Two new institutes were set up, the Institute of Nuclear Power Operations and the Nuclear Safety Analysis Center, for the evaluation and dissemination of information concerning all reactor accidents. The CEGB reported at length at Sizewell on the measures taken to improve the design of the Sizewell B PWR beyond that of the TMI-2 reactor. One other essential difference is the reliance placed on automatic safeguards: the operators on a CEGB station are not required to take any action for 30 minutes following a fault, so that they have time to evaluate the situation they are facing.

The accident at Chernobyl in April 1986 was worse by many orders of magnitude—indeed, it was one of the worst accidents that could happen to a nuclear reactor. The RBMK reactors are large, commercial reactors built only in the USSR. They are graphite-moderated, water-cooled reactors, with fuel pins of uranium oxide clad in zirconium alloy inside zirconium pressure tubes through which passes the cooling water at high pressure. The accident occurred while the operators were attempting to carry out a test at low power. Owing to gross malpractice and a serious design fault, which leads to an RBMK reactor becoming inherently unsafe at low power, a power surge occurred which caused the fuel to overheat and disintegrate. Fragments of fuel at very high temperatures were dispersed in the surrounding water, causing a steam explosion that shattered the top of the reactor. This was probably followed by a hydrogen explosion, due to the escape of hydrogen formed by the reaction of steam with hot zirconium. The reactor core was open to the atmosphere and the exposed hot graphite began to burn. Fires started in the reactor hall and were fought with much heroism. The escape of radioactivity continued for nine days, despite attempts to quench it by dumping 5000 tons of material on top of the reactor; the emission was finally brought to an end by injecting cold nitrogen.

The sequence of events and the cause of the accident have become clear following the release of detailed information by the Russians at a meeting convened by the International Atomic Energy Agency in Vienna in August 1986. The underlying cause was a serious defect in the neutron economy of the RBMK reactors below 20 per cent of full power. This property arises because of the relative weight of two competing effects which control neutron levels. The reactor has a 'positive void coefficient' since, as the fuel temperature increases and the water in the fuel channels changes into steam, more neutrons are available for fission. But as the fuel temperature increases, there is an automatic drop in neutron population owing to the nuclear properties of the fuel itself. The net effect depends on the power level. At normal operating power levels, the fuel temperature effect dominates and the net 'power coefficient' is negative—that is, the fission reaction would tend to slow down as the temperature increases. But at low power levels, the reverse is the case; the power coefficient becomes positive and the reactor becomes unstable. Any increase in power leads to a further increase in power and the design of the shut-down system is poor and too slow in its operation to maintain control of the neutron levels.

The Russians knew this but relied on the operators obeying regulations that dictated that the reactor should never be run continuously at these power levels, a mistake later acknowledged by a senior Russian to be 'a serious psychological blunder'. In the event, the operating team on the night of 25 April 1986, ran the reactor at 6 per cent of full power for two hours, and shut off or circumvented the emergency core cooling system and other protection systems that were warning of incorrect pressure and water levels. They finally ignored a warning that the reactor control rods were incorrectly positioned—they had only eight control rods in the core instead of the required minimum of sixteen—and blocked off another shut-down signal in order to carry out multiple tests. The power level started to rise as they commenced the test and an emergency shut-down was attempted but the mechanisms driving the control and shut-down rods were too slow to prevent a disastrous surge of power a few seconds later (IAEA 1986).

It is clear that the RBMK design suffers from several key faults which contributed to this disaster. The most important has already been mentioned: the reactor has a fast-acting positive power coefficient at power levels below 20 per cent. There is no engineered protection system to prevent operation in such an unstable state. Then the control rods are motored into the reactor, instead of falling under gravity on a trip signal, and there is no secondary system for shut-down in an emergency, as is now established practice elsewhere. In addition, the graphite temperature is very high compared to other graphite-moderated reactors, there is insufficient protection against failure of the pressure tubes, and the reactor containment is inadequate for a water-cooled reactor. It was also obvious that the reactor safety systems could easily be switched off by an operating team on its own initiative.

The above design faults have been recognized by the Russians and the worst of them are being corrected. All control rods will be fitted with limit switches so that they cannot be withdrawn completely, and the equivalent of seventy rods will be kept within the core. Automatic means of preventing operation below 20 per cent power will be added. Longer-term developments include measures to reduce the positive power coefficient and to install a fast-acting shut-down system. Taken together, these measures should be effective in preventing a similar accident in the future (Gittus *et al.* 1988).

As well as design faults, the Chernobyl disaster revealed a faulty safety philosophy and poor standards of safety discipline and training of the operators in the USSR. This too has been admitted. Far too much responsibility for safety was left to the operators and they were able to ignore safety rules and to shut off safeguard systems; further, their training was inadequate and they were left to carry out a test well outside normal operating circumstances without proper supervision. The Russians do not seem to have had an independent organization able to inspect and to check safety standards. These are reports of far-reaching changes in the organization of the nuclear power projects in the USSR (Shabad 1986), and the site director at the time and several senior officials have been prosecuted for criminal negligence.

### The aftermath of Chernobyl

Very large releases of radioactivity resulted from the original steam explosion and persisted for nine days, including 3–6% of the fuel, all the rare gases and 30–50% of the volatile fission products like iodine or caesium. Very high radiation doses were suffered by the firefighters and by other personnel on the site; 203 were diagnosed as suffering from acute radiation syndrome and 31 have died. No member of the population away from the reactor site itself suffered doses large enough to cause acute radiation syndrome. The population of Pripyat, less than 10 km from the site, was first instructed to stay indoors and then evacuated on 27 April; during the next few days 135 000 people were evacuated from an area up to 30 km from the site. Medical support seems to have been very thorough.

The radioactive cloud spread first into northern Europe and then gradually across the whole of eastern and western Europe, reaching the UK by 3 May. Deposition on the ground depended on rainfall, and led to restrictions on the consumption of foods from some areas, including the upland sheep-rearing areas of Britain. The UNSCEAR (1988) report contains an evaluation of the radiation doses received in the first year in all European countries, and also of the collective dose commitment to the world's population. In the first year, average doses following Chernobyl reached 32 per cent of background (0.77 mSv) in Eastern Europe, 11 per cent of background in the USSR and 2–3 per cent of background in the UK. In the long-term, the dose commitment is calculated to be 1.2 mSv in Eastern Europe and 0.15 mSv in Western Europe, with lower values elsewhere. Most of this dose commitment will have been received in thirty years, so that the radiation doses to the populations of Eastern and Western Europe will effectively increase by amounts equivalent to six months and three weeks of background radiation respectively in the next thirty years. These projections are based on past studies of the radioactive fall-out from atmospheric testing of nuclear weapons up to 1980, which was larger. The world-wide collective dose commitment from these tests amounts to an extra three years' exposure to background radiation over many hundreds of years.

While the dose–risk relationship at such low doses is uncertain (p. 95), application of the conventional value would imply an additional 5000–10 000 cases of fatal cancer in the USSR due to the Chernobyl accident over the next ten to thirty years. Most of these would occur in the

European part of the USSR, in a population of some seventy five million people who would suffer some nine million cancer deaths due to other causes in the same period. It is highly unlikely that the radiological impact of this accident will be detectable in mortality statistics (Clarke 1988). In most countries, the extra doses received are comparable to or less than the variation in background dose between different areas.

The Russians must be commended for the energy and resolution which they showed in acting to protect the threatened population and to limit the spread of contamination. Concrete containment has been built around and beneath the reactor. The Chernobyl site was sufficiently decontaminated to allow the other three reactors to be operated in the winter of 1986/87. It seems that they plan to complete the six RBMK reactors which are in the course of construction, though they have said that they will build no more, but instead continue their programme of PWRs, of which they have twenty-seven 950 MW(e) units under construction. Nuclear power is seen as an essential ingredient of the economic development of the USSR and the pressure to make rapid progress may have contributed to the deficiencies in organization and performance which were illustrated starkly at Chernobyl (Marples 1986).

The history of this disaster, by far the worst in the history of nuclear power, will be examined minutely for a long time to come. However, it is not likely that any design changes to the very different types of reactor which operate outside Russia will seem desirable. Reactors with the characteristics of the RBMK reactors would not receive a licence in the UK (Gittus *et al.* 1988). Indeed, the paradoxical result of Chernobyl is to emphasize the importance of the design and operational principles to which all Western countries are working. These principles were not found wanting at Chernobyl; on the contrary, they had been ignored by the reactor designers and flouted by the reactor operators. But one lesson that must not be forgotten is that reactor safety is a matter of international concern. In the aftermath of Chernobyl, member nations of the IAEA signed two conventions, one to guarantee early notification of any nuclear accident that could result in a spread of radioactivity into the environment, and the other to promise prompt assistance to any member nation in the event of a serious nuclear accident. These international arrangements have recently been strengthened by the formation of the World Association of Nuclear Operators by the 130 electrical utilities in the world which operate nuclear power plants. The Charter of this Association obliges members to provide early notification of any significant event in a reactor and to analyse these events to identify possible precursors of serious accidents and to disseminate the lessons learned. The Association will maintain regional offices in Moscow, Tokyo, Paris, and Atlanta, USA.

**Inherently safe reactor designs**

Despite the real advances made in the last 15 years in the methodology of risk assessment and towards the definition of tolerably low levels of risk, some difficulties in basing a safety case on such techniques will persist. An analysis based on probabilistic methods cannot yield guarantees of safety. The procedures of conducting the analysis and sifting the reliability data are complicated and the scrutiny of a safety case must remain a job for an expert. Some doubts concerning completeness—has everything been thought of?—and the possibilities of human failures will remain.

So it is not surprising that many suggestions have been made for the development of reactor systems that may be seen as intrinsically safe, and that are dependent for safe shutting down and cooling on the laws of physics and not on the reliability of engineered systems. Typical suggestions include reducing the power density of the reactor core, relying on passive means of dissipating heat and on automatic shut-down mechanisms if power or temperature increase, and increasingly robust fuel designs to minimize fission product escape.

Some increase in safety may be achieved by modifications to existing designs. The depth of water in which the core of a PWR is submerged in a pressure vessel can be increased and there are various designs for double containment of PWRs and BWRs (Edwards 1985). Then the containment structure can be vented through a filter-bed to reduce the proportion of fission products that would escape into atmosphere following all except explosive pressure rises in the containment. Such a device is being fitted to BWR plants in southern Sweden, and is being considered for the earlier French PWR stations. Typical of much of the new design thinking is an integral design of a pressurized water reactor of 320 MW(e) output in which all the major primary circuit components are contained within a single pressure vessel which has been proposed by a consortium of companies in the UK and the USA. This Safe Integral Reactor design relies on passive safety systems for decay heat removal; the large amount of cooling water above the reactor core allows heat removal by natural convection paths. There is no large-bore external pipework and the large primary coolant inventory leads to a long time being available for remedial measures in the event of a small loss of cooling accident.

Other concepts are even more radical. The 'process inherent ultimate safety reactor', PIUS, designed by ASEA-ATOM of Sweden, consists of a reactor core and a primary cooling system similar to those of a PWR but immersed in a pool of cold, borated water in a large concrete pressure vessel. The concept is that the borated water floods into the reactor core if the primary circulation is disturbed and then provides sufficient cooling by natural circulation for a week, allowing time for other remedial measures (Weinberg and Spiewak 1985). A quite different concept is a modular version of a high-temperature, gas-cooled reactor. This type of reactor uses, as fuel, particles of 20 per cent enriched uranium oxycarbide coated with graphite or silicon carbide which are retentive of fission products up to temperatures of about 1900°C and therefore act as an efficient primary barrier to migration. Once the fission reaction has been quenched, the dimensions of the reactor are small enough to allow for the removal of the decay heat by passive means, by conduction, convection, and radiation to the surroundings (Dean and Simon 1987). A feature of all designs that rely only on passive means of heat removal is that they have lower power ratings than current reactors; the basic modules of both the PIUS and HTGR designs are limited to around 200 MW(e) output.

The very high initial costs of a new design of reactor mean that the introduction of radical new designs will be slow, and confined to circumstances where a long-term market can be expected. Steady improvement of existing designs is more probable in the short term. However, it must be recognized that the true value of modifications or of new concepts can only be appreciated after a rigorous safety analysis has been carried out of the type described in earlier sections of this Chapter.

## Transport of fuel and other radioactive materials

Spent fuel after discharge from reactors is normally stored under water in storage ponds at the reactor sites; an exception in the UK is the site at Wylfa, where the fuel is kept in a dry store, cooled by a current of air. But whatever method of storage is used, spent fuel has to be transported at some time to a central store or to a reprocessing plant. Spent fuel is the most highly radioactive material that has to be transported, though movements of it are by no means the most frequent transfers of radioactive materials; far more journeys are made carrying sources for industrial radiography and for medical purposes.

The NRPB has estimated the radiation exposure resulting from the normal transport of radioactive materials within the UK by road, rail, and air. The most significant source of exposure for workers and the public arises from the transport of isotopes for medical and industrial use; technetium generators for hospital use account for some 50% of the occupational collective dose and other isotopes for a further 35%. About 15% of the occupational exposure can be attributed to transport of materials for the nuclear fuel cycle, including the transport of irradiated nuclear fuel. The total collective dose from all movements of radioactive material is less than 1 man Sv a year, and the maximum annual individual dose received by transport workers in the nuclear industry was 1.2 mSv (Shaw *et al.* 1984).

The environmental impact of normal transport operations is therefore negligibly small. Significant exposures could only occur as a result of accidents. All packages or containers in which radioactive material is carried have to be licensed in the UK by the Department of Transport, and the regulations governing the design of containers conform to international standards set by the IAEA. The most radioactive cargo, the spent fuel, has to be carried in the highest standard of package; the spent fuel 'flasks' must be shown to withstand accident conditions equivalent to:

(1) a drop test from 9 m to an unyielding target, with the flask in the most vulnerable attitude;
(2) an all-engulfing fire at 800°C for 30 minutes;
(3) a drop test from 1 m on to a steel punch;
(4) immersion to a depth of 15 m for 8 hours.

The basic principle is that the containers must be able to withstand accidents without causing an unacceptable spread of radioactivity. Flask designs for this duty are typically cuboid in shape, with external dimensions 2–2.5 m, fabricated either out of 14-inch thick steel or 3.5-inch steel surrounding 7 inches of lead. The gross weight is about 50 tons. The flasks are normally filled with water to aid heat dissipation and closed by a heavy lid bolted on to the main structure. A typical load would be 200 magnox fuel elements or 20 AGR fuel elements, which are larger and emit more heat because of their longer irradiation time.

Considerable studies have been made of the degree to which the IAEA criteria mirror the conditions that might apply in the worst accidents, and of the means by which demonstrations of compliance with the criteria can be achieved. Flask designs are normally tested by the use of replica scale models because of expense and engineering difficulty, and the scaled-down tests are supported by theoretical engineering analysis. Flasks for transporting magnox fuel elements are tested at one-quarter scale. Additionally, the CEGB has conducted two tests on full-scale flasks. A drop test of a full-size magnox flask was carried out at a site in Cheddar in March 1984, from 9 m on to the flask lid edge; the measurements agreed well with extrapolated values from the scale model tests and the flask behaviour was satisfactory, with the release of only 0.4 per cent of the contained water during the actual impact (Morse and Wilkinson 1985). In July 1984 the CEGB organized another test in which a train travelling at 100 m.p.h. crashed into a fuel flask standing on top of a flat railcar placed across the track, which would be one of the worst possible railway accidents. The locomotive was destroyed, coming to rest over 90 m from the point of impact. The fins on the flask suffered only some superficial damage and some bolts were distorted, but the loss of pressure of the contained fluid was insignificant and it is likely that no leakage occurred (Hart et al. 1985).

Spent fuel has been moved for more than 20 years by road, rail, and sea.

In the UK, there have been about 7000 shipments of fuel flasks to Sellafield by rail, amounting to two million miles. The only recorded incidents have been minor derailments at slow speeds in marshalling yards. In addition, spent fuel flasks have been transported for a total of 100 000 flask miles by road. In none of these operations has there been any accident involving damage to a flask or the release of radioactivity (Salmon 1984).

Transport of spent fuel by sea has been carried out without incident for the same length of time; the longest voyages are those between Japan and Cap la Hague or Sellafield. The newest ships used in this traffic are specially designed with additional safety features such as a double-bottom structure and precise navigation and communications equipment; these ships have a better-than-average chance of surviving the normal maritime hazards. The most severe hazard is a collision followed by fire; the probability of such an accident has been calculated to be 1 in 30 000 a year, with a maximum resulting collective dose of 300 man Sv if the accident occurred near a major city. Even in such an improbable accident, the ships' crews are the only people likely to suffer serious consequences (Salmon 1984). Brown (1986) estimated that the collective dose to the UK population resulting from the loss of a leaking flask in deep sea close to the British Isles would be below 0.8 man Sv, which is a very low exposure.

The HSE has carried out a safety assessment of the transport of plutonium as nitrate solution between Dounreay and Sellafield (HSE 1978). An additional safety point that must be watched in that case is the possibility of the generation of internal pressure owing to the radiolytic decomposition of the solution. The Inspectorate concluded that the chance of an accident severe enough to breach the container during road and sea transport in locations where harm might be done to the public was less than 1 in 1 million a year. The risk to an individual would be smaller by at least a factor of ten.

Transport risks were examined closely at the two recent public inquiries. The Sizewell Inquiry concentrated on the risks of transporting spent fuel flasks by rail, arising both from accidents and from malicious damage. The annual probability of a serious accident was thought by the CEGB to be about one in a thousand million, and by consultants to the GLC to be 1 in 250 000. The probability of an actual release of radioactivity would be less than these figures. The Report notes that the 'safety of spent fuel transport gives extremely little justification for anxiety'.

Spent fuel is not carried by air but the plutonium product of reprocessing is and the possible risks were discussed at the Dounreay Inquiry (Brown 1986), and have been reviewed by the Advisory Committee on the Transport of Radioactive Materials (ACTRAM 1988b).) Plutonium would be carried as oxide in strong containers that could only be breached by a high-speed impact on to a very resistant surface such as massive rock. Even such an unlikely accident would not cause a very serious contamination.

ACTRAM concluded that the transfer of plutonium by air was acceptably safe under present conditions.

These very low probability estimates for serious accidents cannot be validated by examination of past statistics since no serious accidents have occurred. An earlier Report from a Study Group of ACTRAM (1988*a*) notes that, partly due to regulatory control, there has never been a serious accident involving the dispersal of radioactive materials during transport in the UK. There were 330 reported incidents in twenty years, out of 750 000 shipments. The major component of both occupational and public doses attributable to accidents arose from the movement of radiography sources; only 1.5 per cent arose from the transport of nuclear fuel cycle materials.

The good safety record and the design and testing procedures adopted for the containers lead to the conclusion that the environmental impact of transporting spent fuel and other radioactive materials will be very low indeed.

## The reprocessing of spent fuel

Spent fuel is reprocessed to recover usable fissile material, uranium for use in thermal rectors and plutonium for use either in fast reactors or in some thermal reactors (p. 103). Metal fuel, as used in magnox reactors, is best reprocessed because of its chemical reactivity. Oxide fuel can be stored for long periods and is technically suitable for direct disposal without reprocessing, but most oxide fuel currently arising in Europe will be reprocessed. The three major plants now operating commercially in the West are situated in Marcoule and La Hague, in France, and in Sellafield in Britain. The annual throughput of metal fuel from magnox reactors is some 1000 t at Sellafield and 600 t at La Hague and Marcoule; 400 t per year of oxide fuel is reprocessed at La Hague, and two new plants, each of 800 t per year are being built there; the new oxide reprocessing plant, THORP, at Sellafield will take 600 t per year or more (RWMAC 1984). There are several smaller experimental plants in Europe and the Dounreay pilot plant which reprocesses fast reactor fuel. The only reprocessing in the USA now is for military requirements.

The first task is to separate the fuel itself from the fuel can and from other parts of the fuel assembly. Magnox fuel cans are taken apart mechanically: the magnox can is stripped of the uranium by forcing the fuel can through a die. In contrast, oxide fuel pins from AGRs or PWRs are chopped into short lengths and the uranium oxide dissolved out by an acid leach. This chemical process leaves less of the fuel on the fuel cladding, thus reducing the activity of the fuel pin wastes.

The fuel material, metal or oxide, is dissolved in hot, strong nitric acid. At that stage, some gaseous and volatile fission products are lost, including carbon-14 as carbon dioxide, krypton-85 and iodine and some of the tritium and ruthenium. The gases are widely dispersed, and the extent to which it may be necessary in the future to remove them from gaseous effluents will depend on the assessments of the importance of the small, but widespread radiation doses that result. Chemical techniques are available to remove carbon dioxide and iodine, and physical processes have been devised to trap krypton if it were deemed to be necessary.

The nitric acid solution is fed into a solvent extraction plant, and separation and purification are effected by contact in successive stages with solvents immiscible with water, and that dissolve preferentially one or other of the required chemical species. This leaves almost all the fission products in the aqueous, acid solution. In the first series of stages of magnox reprocessing at Sellafield, some 99.9 per cent of the uranium and 99.98 per cent of the plutonium are separated from all but 0.5 per cent of the fission products. In a subsequent cycle, uranium is separated from plutonium by extraction into another solvent by means of the addition of ferrous sulphamate to convert the plutonium to a state more soluble in water. The uranium and the plutonium are further purified by separate solvent extraction cycles (Allardice *et al*. 1983).

The details of the separation process have to be adapted to the type of fuel and particularly for the length of time for which the fuel has been irradiated, which determines the concentration of fission products. Highly irradiated fuel contains more fission products, and some of these, metals such as rhodium and palladium, are insoluble in nitric acid. An alloy of these elements has to be removed from the process solution before it goes forward to the solvent extraction plant. This is accomplished by centrifuging the solution; the fission product alloy, after a period of cooling, is added to the fission product wastes. The processing of fast reactor fuel has to be adapted to the increased concentration of plutonium and to the higher radiation levels from long-irradiated fuel, which would cause deleterious chemical changes in the organic solvents unless times of contact were reduced.

The general safety considerations that have to be met in the design and operation of these plants are typical of any chemical plant in which strong acids and flammable solvents are used. There are two additional requirements which determine the design of radioactive reprocessing plants: first, the need to provide adequate shielding and therefore to control all operations remotely while the level of radiation is high and second, the need to guard against any possibility of a sufficient quantity of fissile material accumulating in any part of the plant to form a critical mass, which would lead to rapid nuclear heating, causing local dispersion of material and plant damage. Criticality control is achieved by five methods: limiting the total mass of fissile material in any section of the plant; limiting the volume of any section by design; monitoring concentrations; arranging the plant geometry to facilitate the escape of neutrons; and introducing neutron absorbers such as boron, cadmium, or gadolinium nitrate.

## Plant effluents

All reprocessing plants give rise to a number of liquid waste streams of low activity and these, after treatment, constitute the liquid effluent from the site, together with water from the fuel storage ponds. At Sellafield, low-level liquid wastes from the various sources are finally monitored and, if appropriate, combined for discharge through sea pipelines that extend out to sea for a distance of about 2.5 km beyond the low tide mark. The limits for radioactive discharges are set by the Department of the Environment and the Ministry of Agriculture, Fisheries, and Food, and the considerations guiding these authorizations are discussed below. The liquid discharges are far more important than the gaseous discharges, both because more radioactivity is involved and because the potential effect on the local population is greater.

Discharges from Sellafield increased in the early 1970s mainly due to a delay in reprocessing magnox fuel, which corroded in storage ponds, leading to a release of fission products, especially caesium. The actual discharges of alpha-activity from 1970 to 1988 and projections to 1995 are shown in Fig. 3.6. At the peak, in 1975, discharges from Sellafield were a significant fraction of the radioactivity emitted from all civilian nuclear plants. Although the discharges were below the authorized levels of the time, British Nuclear Fuels (BNFL) decided that action had to be taken to reduce doses to the critical groups. As well as short-term measures, several major new plants were built, namely an entirely new fuel storage and decanning complex for both magnox and AGR fuel; the SIXEP plant, an effluent treatment plant designed primarily to remove caesium and particulate matter from pond discharge liquors by ion exchange methods, and the Salt Evaporator, for the treatment and storage of some of the waste streams from the reprocessing plant. Some other effluent streams are being sent to storage tanks to reduce alpha discharges (Mummery 1985). As a result of these measures, annual discharges have reduced from 9000 TBq to below 100 TBq of beta-gamma emitters by 1988, and from 180 to 2.1 TBq of alpha emitters. Plans for further reductions to meet the ALARA requirement were considered by the company and by the Radioactive Waste Management Advisory Committee (RWMAC 1985). As a result, a further new effluent plant, the Enhanced Actinide Removal Plant (EARP), is to be built to come into operation in 1992; the target is that alpha discharges will be reduced to 0.74 TBq a year and beta discharges below 300 TBq a year; further reductions should occur early next century, with the cessation of magnox reprocessing (RWMAC 1984).

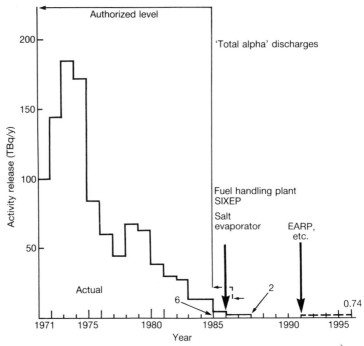

**Fig. 3.6** Actual and projected discharges from Sellafield, 1970–1995 (after Mummery and Anderson 1987).

The historical and projected discharges are compared with those from other plants in Table 3.6; discharges from Marcoule, an inland site, are kept lower than those from the other, coastal, sites. The need to bring the operations at Sellafield into line with international practice was one reason leading to the decision to commission the EARP plant at Sellafield; when it is in operation, discharges of alpha activity from all European plants should be below 1 TBq per year.

**Radiological control**

The limits set upon discharges from these plants are derived from the maximum estimated radiological impact on critical groups of people. Prior to the mid-1970s, the critical group affected by the Sellafield discharges was defined by the consumption of laverbread made from the seaweed collected off the Cumbrian coast, and the· important nuclide was ruthenium-106. However, the increased discharges in the 1970s were largely of Cs-137, which becomes dispersed in the Irish Sea and more distant waters, and which becomes concentrated in the flesh of fish and of shellfish. The critical pathway to man therefore became the consumption

**Table 3.6**
Liquid discharges from reprocessing plants, actual and projected

| Plant | Scale of operation | Date | Liquid discharges (TBq/yr) | |
|---|---|---|---|---|
| | | | 'total alpha' | 'total beta'* |
| Sellafield | 1300 t U/yr magnox | 1983 | 14 | 2480 |
| | 1300 t U/yr magnox | 1987 | 2.2 | 89 |
| | 1400 t U/yr magnox | (1992) | (0.7) | <300 |
| | and    600 t U/yr oxide | | | |
| | 600 t U/yr oxide | (> 2005) | (0.4) | (110) |
| La Hague | 200 t U/yr magnox | 1983 | 0.5 | 1180 |
| | and    250 t U/yr oxide | | | |
| | up to 1600 t U/yr oxide | (1992) | (0.5) | (1100) |
| Marcoule | 400 t U/yr magnox | 1983 | 0.06 | 62 |

*Not counting tritium
*Source*: RWMAC (1984), Hunt (1988)

of local fish and shellfish, and members of a small coastal fishing community local to Sellafield became the critical group. Radiation doses to this group rose to 2 mSv a year, with a contribution resulting from alpha discharges of about 0.2 mSv. Two factors then contributed to a further rise in assessed dose: the regulatory bodies reported an increased consumption of molluscs, such as winkles, by the critical group, increasing from 2 kg a year to the highest value of 16.5 kg a year in 1983, and the NRPB advised that the radiation dose resulting from the ingestion of plutonium should be increased by a factor of five. These changes meant that the critical group became the local mollusc eaters and their annual radiation dose reached 3.5 mSv in 1981.

Since then, the measures taken to reduce discharges have taken effect and the latest figures from the regulatory authorities for public radiation exposure due to the Sellafield discharges are collected in Table 3.7. Few of the critical groups now receive above 0.1 mSv/y and none receive more than 0.5 mSv/y. These figures should fall as the discharges continue to fall, perhaps to 0.1 mSv/y by the end of the century, based on current habits of the critical group (Hughes *et al.* 1984). The regulatory authority's latest report includes the comment that the dose rates that were above the 1 mSv/y level in the recent past have not occurred for long enough for the exposure of the critical group during their lifetimes to have exceeded 1 mSv/year on

**Table 3.7**

Public radiation exposure from discharges of liquid waste from Sellafield, 1987 (Hunt 1988)

| Establishment | Radiation exposure pathway | Critical group | Exposure, mSv/y |
|---|---|---|---|
| Sellafield | Fish and shellfish | Local fishing community | 0.10–0.33* |
| | Handling of fishing gear | Local fishing community | <0.1 |
| | Laverbread eaters | Consumers | <0.01 |
| Springfields | External | Houseboat dwellers | |
| Capenhurst | Shellfish | Local fishing community | <0.1 |
| Chapel Cross | Fish and shellfish + external | Local fishermen | <0.2 |
| Heysham | Fish and shellfish + external | Local fishing community | 0.06–0.12* |
| Hunterson | Fish and shellfish + external | Local fishing community | 0.03 |

*The higher figures were calculated using an enhanced figure for transfer of plutonium through the gut.

average (Hunt 1988). The figures in Table 3.7 include the contribution from the fall-out from Chernobyl, which was mainly caesium.

A large proportion of the alpha-active, actinide compounds is not dispersed into the oceans but remains on the sediments close to the point of discharge. Another pathway to man therefore consists of re-suspension from the sea, with transfer of particulates inland. In order to predict future levels of actinides in the environment, a model of this process has been constructed and tested against the historical record of actinides deposited on land (Howarth and Eggleton 1987). The critical group for this pathway would be agricultural workers who inhale more activity re-suspended from the soil than the average person and who may eat more locally-produced food. The results of this exercise are that the dose from this pathway to a member of the critical group in Seascale reached a peak of 0.035 mSv/y in 1973; this is calculated to reduce to 0.02 mSv/y in 1985 and to 0.016 mSv/y in the year 2000. Exposure falls away rapidly with distance from Sellafield. This pathway is therefore of minor importance compared to the consumption of seafood.

Both beta-gamma and alpha activity in fish and shellfish are monitored around the coasts of Britain and more widely. The resulting collective dose for the UK due to liquid discharges was a maximum of 130 man Sv in 1979, reducing to 70 man Sv in 1984 and to 30 man Sv in 1987, including some contribution from Chernobyl (Hunt 1988). This was the predominant contribution to the collective dose to the UK and European public owing to the operations of the nuclear industry as a whole and it is reducing as the discharges reduce.

These doses are already low and it is instructive to look at the arguments used to justify the expenditure of a further 190 million on the EARP plant at Sellafield, which promises a further reduction of discharged activity by a factor of about four. The dose reduction was estimated to save one or two cancer deaths in the UK over the next 10000 years (RWMAC 1985), and could not be justified on the basis of the cost : benefit criterion usually adopted (p. 97). However, reductions might be necessary to bring the critical group dose below the 0.5 mSv/y target which had been adopted by the company and had been endorsed as a target by the Radioactive Waste Management Advisory Committee. Further, as the committee noted, there are uncertainties in assessing exposures of the public owing to the complicated scientific issues involved, and the general public and political situation of the time required that plants in the UK were brought up to the best standards attained elsewhere. The Advisory Committee therefore endorsed the proposed expenditure but saw no justification for even greater expenditure that would bring little extra benefit.

On the world scale, reprocessing is not the major contribution to the global collective dose from nuclear power because only about 10 per cent of the nuclear fuel used in the world has been reprocessed to date. In order to estimate the possible global impact of a nuclear programme in the more distant future, when the nuclear industry may be making more use of breeder reactors, UNSCEAR (1982) considered the likely effluents from a model oxide fuel reprocessing plant based on the THORP design, and estimated a collective dose which would be 18 per cent of that caused by nuclear power production as a whole if all fuel were reprocessed; the largest item in the total dose was that arising from gaseous discharges from reactors.

## Plant safety

As in the case of reactors, hazard to the public and to the environment can occur because of accidents as well as from normal operation. A number of well-publicized accidents have occurred at reprocessing plants. A pilot plant built to study oxide fuel reprocessing at Sellafield in 1969 was shut down in 1973 after an incident which led to some escape of a volatile compound of ruthenium into the manned area of the plant, apparently caused by the emission of heat from an accumulation of solid fission products; this incident led to a comprehensive review of the reprocessing of highly irradiated oxide fuel (Allardice 1983). The great majority of incidents result in some spread of radioactivity inside the plant, but not outside. One exception was the release of a quantity of solvent associated with high activity solids with the normal low-activity aqueous waste down the discharge line to the sea from Sellafield in early November 1983. The solvent floated on the sea and was blown onshore, leading to the contamination of one of the beaches. The public was advised not to make unnecessary use of the beach until it was shown to be free of patches of high activity associated with particles of high activity debris, though the general level of local radiation dose remained low.

   The total amount of activity released in this incident did not break the authorization for discharge, but the company was successfully prosecuted for breach of the ALARA principle—that all reasonable steps had not been taken to minimize exposure to radiation—and on the grounds that conditions in the site licence concerning operating rules, instructions, and records had not been observed. A number of actions have been taken by BNF plc to prevent a recurrence of similar incidents. In the short term, deficiencies in operating procedures and documentation were remedied. In addition, engineering modifications to the plant were undertaken to reduce reliance on an operator. The direct lines between the processing plant and the liquid effluent discharge plant were blocked. Improved instrumentation to measure radiation levels was attached to the pipes and tanks, and automatic trips were attached to the pumps on the sea discharge lines so that they would stop on signals of abnormal activity levels; better means of transfer of liquid from the sea tanks back to the plant have been installed (Corbet *et al.* 1986).

The Nuclear Installations Inspectorate reported on the general management of safety at Sellafield in 1981 and has more recently carried out a Safety Audit of the complex of reprocessing plants, storage tanks, and effluent plants (HSE 1986). The NII noted the massive investment in new and improved facilities which has been carried out in recent years and which will continue for several years more, but drew attention to the need for continued maintenance and attention to some of the older plant, including buildings which are closed but not decontaminated. Specific recommendations were made for improvement to the magnox fuel reprocessing plant, and the company was required to submit a revised safety case for its continued operation. All incidents involving any spread of radioactivity within the site from 1979 to 1986 were analysed and five incidents which occurred in 1986 were reviewed carefully. The probability of any incident causing a significant hazard to the public is considered to be remote. In several places in the report, the HSE draws attention to the difficulty caused to the management of the site by the current lack of national agreement on routes for the disposal of solid radioactive wastes, a matter to which we return in a later section (p. 167). The HSE's report also notes that the potential hazards to the public from reprocessing plants are less than arise from nuclear power reactors, because of the large amount of stored energy and relatively rapid response to abnormal conditions characteristic of a reactor; processes in chemical plants are slower acting, and any failure should be detected before it reaches significant proportions.

Reprocessing operations have been the subject of two recent public inquiries in this country: the Windscale Inquiry in 1977 on the application for outline planning permission to build the oxide fuel reprocessing plant, THORP, at Sellafield (Parker 1978), and the Inquiry into the application by the UKAEA and British Nuclear Fuels plc (BNFL) for outline planning consent for a fast reactor fuel reprocessing plant at Dounreay in 1986. Mr Justice Parker's recommendation that THORP should go ahead was endorsed after two parliamentary debates. THORP is now being constructed. The final report from the Dounreay Public Inquiry has not yet been issued, and even if outline planning permission is granted, the construction of a European Demonstration Reprocessing Plant will depend on agreement being reached on the location of this plant by the European partners to the fast reactor collaboration agreement. These public occasions have fulfilled something of the same functions for reprocessing plants as did the Sizewell Inquiry for nuclear reactor construction—that is, they have provided a forum for a wide-ranging examination of the operation of such plants, of the safety philosophy adopted and of the means of monitoring and of controlling any environmental effects. Much time has been spent in all these inquiries in going over the basic standards of radiological safety to which plant operators must conform, and to hearing objections to them from various witnesses wishing to argue that standards should be raised; since these are national standards, common to all nuclear plant, it seems regrettable that there is no other forum in which national debate about them can take place. More relevant to the specific reprocessing plant issues is the detailed examination of critical pathways, the definition of critical groups and the examination of their habits, and of the monitoring techniques and schedules adopted so that radiation doses can be estimated. The assumption that protection of people to an adequate standard will automatically protect local ecologies from harm has also been challenged, but generally upheld.

In addition, the risks of accidents also have to be assessed. Mr Justice Parker considered that the three possible sources of major releases from Sellafield were the fuel element cooling ponds, the plutonium stores, and the tanks of highly active liquid wastes. These last contain the highest concentration of activity and have to be kept cool by pumping water through cooling coils; redundant cooling coils and electrical supplies are available. If the supply of cooling water fails, then eventually the solution will boil, releasing some radioactivity; it was calculated that it would take between 9 and 31 hours for the liquid to reach boiling point and 2.5 to 8 days before the tank boiled dry. The Inspector accepted that it was highly likely that remedial action could be taken in that time. Similar calculations concerning the safety of highly active waste tanks were also presented at the Dounreay Inquiry into the EDRP proposal. In that inquiry, an outline probabilistic risk assessment was presented, in which the consequences of the plant being disrupted by external events such as an aircraft crash, fire and ground movement had been estimated. The safety criteria were similar to those adopted for reactor accidents, namely, a maximum risk of death of one in ten million per year to the most exposed individual, and there was considerable argument about the definition of societal risk targets. However, it should be simpler to establish that these risk targets can be met in a reprocessing plant than in a nuclear reactor, since a plant does not contain such a concentrated source of energy capable in the worst circumstances of releasing large amounts of radioactivity in a short time.

## The disposal of radioactive wastes

Gaseous and liquid discharges from nuclear plants have been dealt with in the preceding sections of this chapter. Such discharges are responsible for all the immediate radiological effects of nuclear power, but they account for a very small proportion of the radioactivity produced, nearly all of which will arise as solid wastes of one sort or another. It is an important safety principle that all wastes stored as liquids or slurries should be converted into a solid form, to reduce significantly any possibility of migration into the environment. Solid wastes are easily stored behind appropriate shielding, though consideration has to be given to the need for ventilation if the radiation field is intense enough to cause some gases to be produced by the radiolysis of water or of organic compounds. But storage cannot be seen as a permanent solution for the long term, and all countries with nuclear power programmes are working towards the establishment of permanent repositories for all nuclear wastes; several have made faster progress towards this end than we have in Britain.

The principle which has guided policy in this country was enunciated by the Royal Commission on Environmental Pollution (RCEP) in its Sixth Report (1976)—that methods of disposal should be developed for which safety could be guaranteed beyond reasonable doubt for the foreseeable future without the need for continued human surveillance. This principle must be regarded as an ethical one, based on long-term concern for the environment and the consideration that we should not leave an intractable problem to later generations. While it is completely fair to require the nuclear industry to demonstrate that this principle can be met, the time-scale for the implementation of this policy and the construction and operation of permanent repositories is a political decision, because of the political need to satisfy public opinion that strict safety criteria can actually be met.

The evolution of safety criteria for repositories has proceeded along broadly similar lines in many countries; the guidance issued by the ICRP is that it should be demonstrated that no individual will suffer in the future an annual dose from waste disposal greater than would be allowed under present regulations, and that the ALARA principle should apply. There is also a general consensus that radiation exposure resulting from a nuclear waste repository should be a small fraction of that due to natural background. Most countries which set quantitative dose limits do so for the maximum individual dose likely to occur; the only exception is the USA, which adds a collective dose limit as well (Parker et al. 1987). In the UK, the radiological criteria that would have to be satisfied before waste disposal could be authorized are that the ensuing risk to any individual from one repository should be less than that corresponding to a maximum dose of 0.1 mSv a year, which is one twentieth of the average background and much less than the variation of natural background across the country (Department of the Environment 1984). It is a more stringent condition than the IAEA guidelines would imply.

Proof of safety to such a standard without time limit is an exacting intellectual challenge. All countries have adopted the same overall strategy; solid wastes will be packaged in a durable form and buried in carefully chosen locations, perhaps with the addition of further engineered barriers to prolong containment and isolation. The techniques adopted depend on the local geology and on the radioactive content of particular types of waste. Wastes containing mainly shortlived species can be disposed of in comparatively shallow locations, while long-lived wastes must be placed remote from the biosphere, either deep underground or under the sea bed. In all locations, once a repository has been closed, there is no hazard from direct radiation because of the shielding value of the cover; the only possibility of radioactivity getting back to man is through contact with water, which may leach some of the radioactive species out of the waste, or through accidental intrusion.

The proof of safety that is required therefore consists of a calculation of the maximum dose that could accrue to an individual, or to a critical group of the population, because of a solution of some of the radionuclides in the buried waste reaching the biosphere. The steps of such a calculation are illustrated in Fig. 3.7. It is first necessary to estimate the lifetime of any container around the waste, and then the rate of leaching out of any soluble radionuclides. The next step is to calculate the degree to which specific radionuclides can be retained by absorption or chemical reaction on any secondary barriers placed round the waste containers. Any concentration that does escape becomes the source term for an estimation of the rate at which the radionuclides could permeate the geological barrier and reach the biosphere. The final step is an estimation of the possible radiation doses to human beings that could result from any pathway through the biosphere, a calculation which is similar to that carried out for effluents. The total sequence illustrated in Fig. 3.7 is characteristic of 'normal' conditions—i.e. on the assumption of no sudden change in the

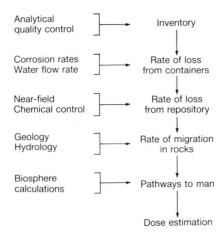

Sequence of safety case

| Analytical quality control | → Inventory |
| Corrosion rates Water flow rate | Rate of loss from containers |
| Near-field Chemical control | Rate of loss from repository |
| Geology Hydrology | Rate of migration in rocks |
| Biosphere calculations | Pathways to man |

Dose estimation

**Fig. 3.7**  Sequence of the calculations made to estimate radiological risks arising from buried radioactive waste.

environment. It is also necessary to carry out a probabilistic estimation of the possible effects of abnormal disruption, either to the hydrogeological conditions due to climatic change or to geological movements, or to the repository due to human intrusion. Because such changes become more likely in the distant future, appreciable quantities of long-lived species should not be deposited in shallow repositories.

All these calculations require an extrapolation of results for long periods of time, and the rates of permeation through geological strata can only be calculated after making geological and hydrogeological measurements on actual or typical sites. Proposed programmes of geological exploration have been hindered in many countries, and particularly in Britain, because of local public hostility that has led to political difficulties. As a result, only very limited geological studies have been carried out in the UK on specific sites, though much more has been done elsewhere. Public unease about waste disposal has led to considerable programmes of research and of explanation being mounted in many countries to support the type of safety analysis that has been sketched above. New organizations have been established to accomplish this task—e.g. ANDRA in France, NAGRA in Switzerland, the SKB in Sweden, and Nirex in Britain. It seems that the actual procedures adopted for site selection and verification are as important as the details of the technical safety case from the point of view of public confidence in the results.

## Types and quantities of solid radioactive wastes

By far the greatest quantity of radionuclides occur in concentrated form in the spent fuel itself, either as fission products or as actinide elements formed as a result of neutron capture reactions. Solid radioactive wastes are conveniently characterized by their specific activity—that is, by the radioactive content per unit weight or unit volume, since this is an approximate measure of the techniques that have to be adopted for safe handling. The broad categories adopted in the UK are defined as follows:

1. *High-level (heat generating) waste (HLW):* HLW wastes are the highly radioactive liquids containing the bulk of the fission products and actinides separated during the chemical reprocessing of irradiation fuel, and solid waste-forms such as glasses formed from them. The radioactivity is so concentrated that heat emission is significant and artificial cooling is required for a time. Spent fuel is regarded as HLW if it is treated as waste.

2. *Low-level wastes (LLW):* LLW wastes are those that do not require shieding during normal handling and transportation, defined in the UK as having less than 4 GBq/t $\alpha$* and 12 GBq/t $\beta\gamma$, which would normally lead to

---

* GBq = giga bequerel, $10^9$ Bq, or 1/1000 of 1 TBq.
  KBq = kilo bequerel, $10^3$ Bq.

a maximum surface dose rate of less than 2 mGy/hour, the present limit for unshielded movement under the transportation regulations. The term 'LLW' is not applied to wastes of such low activity that they may be disposed of with normal household refuse; the lower limit is defined as material containing less than 400 KBq* of $\beta\gamma$ activity.

3. *Intermediate level wastes (ILW):* ILW wastes are those with specific activities greater than the maximum for LLW but low enough so that heat emission need not be taken into account in the design of stores or repositories. A large number of different types of waste falls into this category, and actual specific activities range up to 2 TBq/m$^3$ of $\alpha$ and up to 800 TBq/m$^3$ of $\beta\gamma$; an average is about 40 TBq/m$^3$ or 1/250 of the specific activity of HLW. Nevertheless, shielding is required during handling and transportation.

Broadly similar definitions are used in other countries, though quantitative limits differ. This categorization is not a complete guide to disposal options, since consideration has to be given to differences in the half-life and toxicity of individual components as well as to specific activity when a detailed case for disposal is made. High-level wastes are produced only at reprocessing plants. Waste streams classified as intermediate level occur at reactor sites and at reprocessing plants, and include sludges from fuel storage ponds, ion exchange materials from water treatment plants, miscellaneous reactor components, fuel cladding, flocs, and ion exchange materials from effluent treatment plants. Low-level wastes are generated at every location where radioactive materials are handled. They arise from fuel enrichment and fabrication plants, reactor sites, reprocessing plants, research laboratories, hospitals, and industrial users of radioactive sources. These wastes are a mixed bag of contaminated clothing, glassware, plastics and bits of old equipment, the garbage of the nuclear industry.

The total radioactive content of the wastes from a nuclear power programme is determined simply by the amount of electricity generated, since the dominant influence is the number of fission reactions that have taken place. But the chemical nature and quantities of waste depend on the type of fuel and the type of reactor used, and on the decision whether or not to reprocess the spent fuel. To illustrate this point, Table 3.8 is a summary of the waste arisings per Gw(e) of installed capacity from a typical magnox, PWR, and fast reactor, assuming a twenty five year life for the magnox reactor, forty years for the PWR, and thirty years for the fast reactor. The much larger volume of waste generated by the magnox reactor is accounted for by the lower efficiency of fuel use of a natural uranium compared to an enriched fuel, leading to a larger through-put of the reprocessing plant for the same amount of electricity generated. Waste arising from reprocessing accounts for about half the total volumes of waste, and a higher proportion in the case of magnox reactors. However, if

**Table 3.8**

Lifetime solid waste arisings from magnox, PWR, and fast reactors

Packaged volume (m³) per GW(e) installed capacity

| | HLW | | | ILW | | | LLW | | |
|---|---|---|---|---|---|---|---|---|---|
| | Magnox | PWR | FR | Magnox[1] | PWR | FR | Magnox[1] | PWR | FR |
| Fuel fabrication | | | | — | | 1,200 | — | (3,000)[2] | 3,200 |
| Reactor operation | | | | 1,300 | 2,000 | (330) | 7,500 | 6,700 | 5,400 |
| Fuel reprocessing | 130 | 136 | 93 | 21,250 | 1,500 | 2,000 | 200,000 | 22,000 | 6,000 |
| Decommissioning | | | | 20,000 | 1,000 | (1,000) | 50,000 | 10,000 | (10,000) |
| Totals | 130 | 136 | 93 | 42,550 | 4,500 | 4,530 | 257,500 | 41,700 | 24,600 |

*Source*: RWMAC, 6th Annual Report (1985). Figures in parentheses were estimates.
[1] R. Flowers, CEGB P.21 Evidence to Sizewell B Public Inquiry (1982).
[2] International Fuel Cycle Evaluation INFCE/PC/2/7 (IAEA, Vienna, 1980).

the uranium mill tailings are considered as additional low-level wastes generated by the thermal reactor fuel cycles (see Fig. 3.1), then their volumes would be much larger than that of any other category of waste. An addition to the decommissioning wastes for the fast reactor would be the disposal of about 4000 tons of sodium, containing perhaps 75 TBq of caesium-137, which would probably be converted into sodium carbonate before disposal.

Recent estimates (Nirex 1987) of waste arisings in the UK from now to the year 2030 total 1.5 million $m^3$ of LLW and 250 000 $m^3$ of ILW from civil sources. Apart from the continuing use of Drigg for much of the LLW, one deep repository should be sufficient for fifty years or more. All categories of LLW and ILW can be disposed of when suitable sites are found and proved. The disposal of HLW will be delayed to ease the engineering problems caused by the high initial rate of heat emission (see p. 156).

## Packaging and containers

The objective of packaging radioactive wastes is to facilitate safe handling, transport, and storage. Most low-level wastes are simply packed into standard 200-litre steel drums, sometimes after compacting to reduce volumes. More active wastes are incorporated into durable solids to minimize storage and transport risks. There is the additional bonus that these solid waste forms will provide a valuable first barrier to the leaching action of water when the wastes are finally disposed of.

Intermediate level wastes, which may contain long-lived radionuclides, but which emit a negligible level of heat owing to radioactive decay, are usually cast into monolithic blocks to minimize leakage from the container and to ensure that the package will retain its shape and strength during any conceivable period of storage. The waste form should be chemically stable and resistant to radiation damage. The package should be suitable for transportation and therefore non-flammable and resistant to immersion in water. Intermediate level wastes include a wide variety of different chemical substances, including inorganic and organic ion exchange materials, inorganic sludges from fuel storage ponds, and fuel cladding and other components of fuel assemblies. The waste form chosen for each waste stream must be chemically compatible with its contents, and should also be suitable for the eventual disposal option chosen for the wastes. Most work has been done using concrete, bitumen, and thermosetting resins. Any organic material carries a slightly enhanced fire risk and organics are subject to radiation damage in high radiation fields. Concrete is free of these objections but is a more porous material than cast resins, with consequent poorer resistance to leaching.

British Nuclear Fuels plc has developed a strategy for encapsulation of all the ILW stored on the site at Sellafield, and plans for an encapsulation facility large enough to cope also with arisings of fresh wastes in the future. The main strategy is based on a cement-based matrix although facilities for introducing other encapsulating materials can be included. Complete mixing of the cement grout or mix with the waste in a 500-litre stainless steel drum is ensured by vibro-compaction after the addition of a cement grout or by the use of an internal paddle which can be sacrificed. After the cement has set, a layer of grout will be run on to the cemented waste to form an inactive cap and a stainless steel lid will be fitted. Only alkali elements such as caesium are leached at any significant rate from cements; the leach rates for actinides are usually below the detection limit and the leach rates for caesium can be reduced by appropriate choice of the cement mix (Elsden and Heafield 1985). Special care has to be taken of the packaging of magnox fuel cladding wastes. The magnesium alloy corrodes on storage under water, becoming mainly oxide, but any bare metal could react with the residual water in the concrete, with the evolution of hydrogen gas which could cause the concrete mass to crack sometime in the future. The cement mix used and the temperature reached during the concreting process have to be carefully controlled to minimize this possibility.

Greater demands have to be met by the encapsulation medium used for the high-level wastes; the material will be subjected to a radiation dose about a thousand times larger than from intermediate-level wastes and it will also have to withstand a considerable evolution of heat for many tens of years. The desirable properties of a waste form for HLW therefore include the following: high strength and durability for long periods; good capacity to incorporate and retain a high proportion of the elements in high-level waste; high resistance to radiation damage; good thermal stability and high thermal conductivity; high resistance to the action of groundwaters.

All countries which have developed a process for the immobilization of high-level wastes have used borosilicate glasses, at least for a first generation process. Glasses are durable materials whose composition can be varied widely to optimize the properties desired. That borosilicate glasses can be used successfully to immobilize highly active waste was demonstrated on a pilot plant scale in Harwell in 1964, and plants have operated in several countries since then. The largest commercial operation has been carried out at Marcoule, in France, where over 1200 $m^3$ of high-level waste solution has been processed, producing over a thousand glass blocks. Based on this successful experience, two larger vitrification plants are being designed for the main French reprocessing plant at La Hague, and a similar plant is being built at Sellafield, due to come into production in 1990.

The French process uses two stages: the acid fission product solution is first fed through a furnace to drive off the liquid and to turn the fission products into a mixture of oxides; the mixture is then fed to a melting vessel together with the requisite amount of solid glass; periodically the molten, active glass is run into steel containers. In one variety of this process, a single- instead of a two-stage plant is used, and in another, the glass product is converted into glass beads which are finally dispersed into a lead matrix. The glass blocks have to be kept artificially cool to prevent the centre temperature rising to temperatures at which the glass would crystallize. This is accomplished by cooling with a current of air, either by forced convection or by natural convection; the air is monitored before discharge through a tall chimney.

The glass blocks are not homogeneous. Most of the fission product elements dissolve in the glass structure—which was one reason for the choice of glass as a waste form—but some do not. Elements such as ruthenium, palladium, rhodium, and tellurium are found dispersed as particulates in the glass matrix; this does not seem to be deleterious. Little crystallization (devitrification) of the glasses should take place if the maximum temperature encountered during storage or disposal is kept below 500°C. Also the radiation stability of the glasses seems satisfactory. The most damaging form of radiation is alpha-particles, since these have sufficient energy to displace atoms from their normal sites in the glass structure. The effects of long periods of alpha-irradiation have been simulated by incorporating, in the glass, isotopes such as Pu-238 or Cm-244, which are alpha-emitters of short half-life. It is possible to subject the glasses in a few years to the irradiation doses which they would receive in practice in 1000 to 100 000 years. Density changes up to 1 per cent occur, with some microcracking, but the physical integrity of the waste glasses is not impaired and the rates of solution in water do not alter by more than a factor of two or three (Mendel 1986).

Other waste forms as well as borosilicate glasses have been developed, though none are used as yet on the industrial scale. An interesting example is the range of 'SYNROC' materials based on the composition of natural minerals that can accommodate a wide variety of elements in their crystal lattices. 'SYNROCS' are based on titanate ceramics and typically consist of a mixture of hollandite, zirconolite, and perovskite phases with some rutile, titanium dioxide, and titanium metal added to preserve reducing conditions during manufacture. Elements such as caesium and barium enter the hollandites; uranium and some actinides enter into zirconolite; the perovskites take the lanthanum and cerium and other actinides, and noble metals like ruthenium and palladium form alloys with titanium metal. To produce SYNROC, an appropriate mixture of titanium and other oxides is added to the solution of high-level waste and the slurry dried and calcined in a reducing atmosphere in a rotating kiln. The resulting power is then converted into fully dense monoliths by hot-pressing at about 1150°C, during which process the final mineral phases are formed. The resulting blocks of SYNROC are more resistant to reaction with water than the borosilicate glasses and are stable in water up to 600°C, whereas glasses disintegrate above 300°C (Ringwood and Kelly 1986). SYNROC might therefore be suitable for disposal in deeper geological strata, where the ambient temperature is higher, or possibly for disposal after shorter cooling times, if that is seen to be desirable.

The packaging for direct disposal of spent oxide fuel without reprocessing introduces quite different considerations. The aim in that case must be to contain the fuel pins in a strong, durable container, with as little handling as possible. For example, the Swedish SKBF Company proposes to use a canister of solid copper, perhaps 0.8 m diameter and 4.5 m tall, with a wall thickness up to 10 cm. The spent fuel would be placed in a canister and the voids between the fuel pins filled with molten lead; a copper lid would be welded in place. Alternatively, the voids in the canister could be filled with copper powder, the canister closed and then subjected to high pressure at 500°C to compact the powder to a single, homogeneous body (KBS 1978).

The duty expected of these various forms of waste package and estimates of their behaviour under repository conditions are described in the later paragraphs, after some discussion of the design and siting of repositories.

## Repositories for high-level waste

High-level waste must eventually be disposed of in a situation remote from the biosphere, and the optimum conditions are in deep, stable geological strata through which the movement of groundwater is known to be slow. Much exploratory work has been carried out in various media, for example: salt deposits—studied in France, West Germany, Netherlands, USA; clays—in Belgium, Italy, France; hard rocks—in Canada, Finland, Japan, France, Sweden, Switzerland, and USA; sedimentary rocks—in Switzerland and the USA.

Only limited geological work has been carried out in Britain, owing to the political difficulty of obtaining permission to carry out test drillings on specific sites; one granite area has been drilled in Caithness and one borehole sunk into a deep clay layer underneath Harwell. In addition, the UK has participated in an international effort to explore the feasibility of using the deep ocean sediments. More exotic solutions have been proposed, including firing the wastes into space or using the Antarctic ice cap, but such measures do not seem to be necessary and have made little progress.

Because of the high rate of heat emission and the small volumes of waste concerned, a period of storage of HLW with artificial cooling is advantageous. The choice of optimum cooling time depends on the maximum temperatures that can be tolerated both in the waste form itself and in the surrounding medium after disposal has taken place, and also on the political perceptions of safety. In some cases, a prolonged period of storage would be seen to confer added confidence, since inspection and monitoring can be carried out easily, while in other circumstances a demonstration of ultimate disposal as early as possible may be perceived to be necessary.

The maximum temperature attained in a repository is a matter of repository design; it depends not only on the time of cooling and the resulting heat output of a waste canister but also on the thermal conductivity of the rock and on the size and spacing of the canisters. As an illustration, Fig. 3.8 shows the temperature profile that would result in a repository constructed in granite, with waste blocks containing 15 per cent of HLW placed 20 metres apart, as a function of time of cooling before emplacement. In this example, fifty years' storage would give a maximum temperature of under 150°C, which would fall after 100 years, reaching ambient temperatures after 1000 years; storage for 100 years before emplacement would lead to temperatures no higher than 80°C (Hodgkinson *et al.* 1983). The spacing of the waste packages clearly determines the volume of a repository. If standard glass blocks are placed 20 metres apart—an entirely arbitrary assumption—the high-level waste from a nuclear programme generating a total of 350 GW(e), which is more than

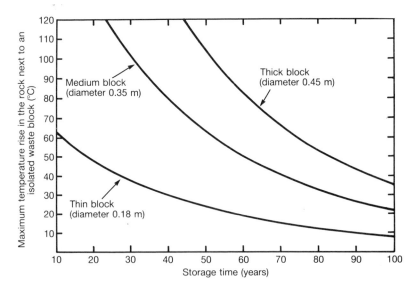

**Fig. 3.8** Maximum temperature rise in rock next to an isolated block of vitrified high-level waste as a function of the pre-disposal storage time. The diameters of the blocks are assumed to be (a) 0.45 m; (b) 0.35 m; (c) 0.18 m.

will be generated in Britain this century, would eventually occupy a repository of perhaps 30 million m$^3$; for example, an area of a half of a square kilometre, with canisters in shafts 60 m deep.

The first requirement in the search for a suitable site is to identify a coherent rock structure of suitable volume. The next is to develop sufficient knowledge of the hydrogeology of the area so that the rate of transit of groundwater from a repository to the surface or to the nearest aquifer can be calculated. There are significant differences between the data required in the rock types being investigated. Salt caverns are dry and tests concentrate on the heat dissipation around the buried waste and the mechanical changes that might occur owing to the construction of a repository, but the consequences of water intrusion, however unlikely, must be evaluated. Clay-like layers contain water which flows through them very slowly; the main questions are the stability of the clays after the introduction of the wastes, since clays tend to crack if they dry out. Crystalline rock formations are subject to fracture, and the fracture pattern will define the path of groundwater movement; fracture characteristics and the complicated flow patterns that result are therefore the most important data.

A large number of different designs of repositories have been proposed. Provided that the geological conditions are acceptable, construction techniques are available and tested for hard rocks and salt deposits (Godfrey et al. 1985). Most designs for hard rocks consist of a network of galleries reached by a shaft or by an inclined tunnel to the surface. The wastes would be placed in the galleries or in holes drilled down from them. Larger caverns could be constructed if necessary. There are also designs for deep-hole repositories, in which cylinders of waste would be placed in drill holes of perhaps 0.75 m diameter, which might be drilled down to 2500 m depth in suitable locations. A conceptual design for a repository in a deep clay formation consists of a series of circular cross-section tunnels, 3.5 m in diameter, with two longer access shafts and ventilation shafts. The ground freezing technique has been used successfully for excavation of a horizontal gallery at a depth of 220 m in the Boom clay at Mol in Belgium.

Whatever rock type is chosen, any repository design will have to satisfy the safety criteria before operation can be licensed; the efficacy of the barriers formed by the container, by the engineering barriers incorporated in the repository (the 'near-field') and by the geological strata between the repository and the surface (the 'far-field') must be proved. The duties for which can barrier should be designed in those cases where groundwater may be presumed to be present—i.e. in hard rock or clay-like deposits—can be best illustrated by considering the timescale over which the various constituents of the waste will decay.

The radioactivity of the high-level waste from reprocessed PWR fuel, with the contributions from the most important constituents, is plotted as a

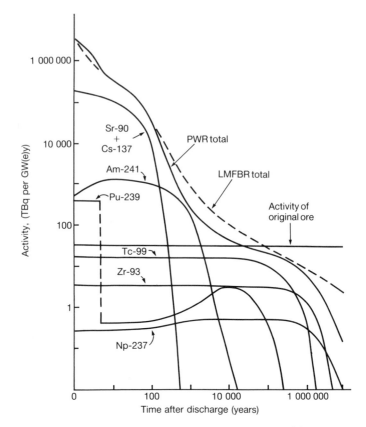

**Fig. 3.9** The radioactivity of high-level waste from 1 GW(e) year reprocessed PWR fuel as a function of time, with the contributions from some individual radionuclides and the activity of the amount of natural uranium ore required to fuel the reactor for a year. The total decay curve for high-level waste from a liquid metal-cooled fast reactor is included for a comparison (Flowers *et al.* 1986)

function of time in Fig. 3.9. The activity figures relate to the high-level waste produced from one gigawatt-year of generation. Reprocessing the fuel is assumed to take place at five years after discharge and to remove all but 0.1 per cent of the uranium and plutonium in the fuel; for convenience, only one plutonium isotope—Pu-239—is shown. The total activity from the high-level waste from 1 GW(e)y of generation in a fast reactor fuelled with plutonium is also plotted in Fig. 3.9, for comparison. In fact, the use of a plutonium fuel makes relatively little difference to the overall decay curve once all but 0.1 per cent of the plutonium is removed with the uranium on reprocessing.

It is apparent from Fig. 3.9 that the most important nuclides for a few hundred years are isotopes of caesium and strontium; when these have decayed away, most of the activity for 100 000 years arises from americium, technetium, and plutonium; at longer times, only long-lived isotopes such as Np-237 and Tc-99 remain. But by this period the activity has decreased by a factor of nearly a million and the total activity left is less than that of the original uranium ore; a line showing the activity of the 200 tons of natural uranium required to provide 1 GW(e)y of electricity from a PWR is shown in the Figure (Flowers et al. 1986).

It is against this background that the performance of the sequence of barriers between the waste and the biosphere has to be judged. The first barrier is provided by the container. A wide variety of alloys have been tested as containers, including stainless steels and titanium-based alloys, and much work in the UK has been carried out on carbon steels. Several of these show such low general corrosion rates that lifetimes of thousands of years under repository conditions would be predicted. Attention has therefore turned to the possibilities of localized corrosion which could cause some penetration of a canister under the chemical and radiation conditions in a repository. This 'pitting' corrosion, as it is called, is more difficult to predict since a statistical spread of spits of different depths tends to form, but again it seems that canisters can be designed to prevent any penetration for 500 years or more (Day et al. 1985).

If the canister is breached, the next barrier to the dissolution of radionuclides is afforded by the slow dissolution of the wasteform itself. The borosilicate glasses used are quite resistant materials, showing a reaction rate in flowing water equal to that of basaltic rocks, which would lead to dissolution of the glass block in about 5000 years if the water penetrated all the cracks in the glass. However, in an underground repository, the rate of dissolution is limited by the very slow flow of groundwater, since the water is almost stagnant and will tend to become saturated (Hughes et al. 1983); the effective rate of solution may then be lower by a factor of 10 000 or more. Even lower rates of loss would be predicted for ceramic waste forms such as SYNROC.

So the combined barrier of canister and wasteform should prevent any radionuclides appearing in the groundwater for at least a thousand years, by which time the activity due to Sr-90 and Cs-137 will have decayed away. This is important, since caesium and strontium are likely to be comparatively mobile species, not strongly absorbed by the materials of repositories or on rocks. The next barrier consists of the surrounding material which is used to backfill the repository, which can be chosen to provide physical and chemical conditions to impede the migration of radionuclides away from the repository itself. Packing materials that have been proposed include impermeable clays such as bentonite, which swell in water to seal crevices and which reduce still further the rate of groundwater movement through the repository, and the use of cement both as a packing and sealing material. Cement has the additional valuable property of maintaining the local conditions alkaline, and if the repository also contains sufficient iron and steel, the local chemistry will be both alkaline and reducing. Under such conditions, elements such as plutonium, americium, and technetium are very insoluble, and will not be dissolved in the groundwater. In addition, both clays and cements have a good capacity of absorbing many elements, still further reducing their concentrations in solution. This type of chemical control does not depend upon any of the constituents retaining their original shape; it is simply the presence in the repository of the correct materials that matters. Under conditions of low groundwater flow, a cement-based repository should be capable of maintaining alkaline conditions for up to one million years (Flowers and Chapman 1986). It is then probable that the actinide elements, with the possible exception of the very long-lived Np-237, will never leave the repository environment in significant concentrations.

The final barrier to radionuclide migration is the 'geological barrier' provided by the properties of the surrounding rocks. This has a dual purpose: it is always in place to act as an insurance against some unforeseen circumstance that may reduce the efficacy of the local barriers, and it is the final barrier delaying the very long-lived species reaching the biosphere. The geological survey work that is going on in many countries to understand the factors controlling groundwater flow in hard rock and in clay deposits has already been mentioned. The flow equations that result from this work are tested in underground laboratories that have been constructed in many locations—for example in the USA, Canada, and Sweden. The use of natural analogues, such as occur in thorium deposits or in uranium mines, is one method of extending validation over long timescales (Rae and Black 1986; Department of the Environment 1987). The most famous example is the study of the debris from a natural reactor that was formed hundreds of millions of years ago in the Oklo region, at a time when the concentration of U-235 in uranium ores was much higher than it is today. The actinides and insoluble fission products can still be traced close to the initial location. The underground experiments are also being used to test different techniques of sealing a repository, for the seal should be as sound as the host rocks.

An early, but thorough study of a complete repository system in hard rock was that made by the Swedish KBS project of the disposal of vitrified high-level waste in the Swedish bedrock (KBS 1978). On the basis of corrosion tests, penetration of the canister of steel and titanium in 1000 years was assumed, with complete accessibility of the glass contents in 6000 years. The lifetime of the glass blocks in the groundwater was estimated to lie between 30 000 and 3 million years. Groundwater transit times from a repository to a lake or well were taken to be 400 years on very conservative assumptions; the actual transit times are said to be more than 4000 years. The highest estimated doses were found to arise from drinking water from a well in the vicinity of the repository; the maximum dose was 0.13 mSv/y occurring after 200 000 years, and a more likely result was said to be a hundred times less than that.

Predictions of the effectiveness of the geological barrier over such long timescales have to take account of the major changes in climate and geology which may occur. Possible changes in sea-level in the Baltic Sea were taken account of in the Swedish study referred to above. For example, the next glaciation may occur in 5000–20 000 years, with consequent stress on surface rocks, enhanced erosion and changes in sea-level. Adverse consequences on a repository may be avoided partly by careful siting and partly by constructing the repository at an adequate depth. Radioactive decay will have reduced activity levels considerably by that time, so that the erosion of the surface rocks could actually release more radioactivity than would come form the residual contents of a repository (Roberts 1979).

The safety case that has to be made for disposal in thick salt deposits is entirely different. The surroundings will be dry and no spread of radioactivity is credible unless water enters the repository at some future date; rock salt is relatively very soluble in water and water intrusion must therefore be shown to be very improbable. It is important to be able to predict the geological stability for many thousands of years and to show that human intrusion is unlikely. A number of underground tests and demonstration experiments are in progress; it seems probable that long-term safety can be proved in a number of locations where extensive salt domes exist (Kühn 1986). The EEC has been conducting for many years a comparative study of the safety of high-level waste disposal in four locations; a salt dome, deep clay, granite, and deep sea sediments. It seems likely that all these possibilities will be shown to be acceptably safe.

### Direct disposal of spent fuel

In principle, spent fuel can be disposed of in any repository suitable for high-level wastes. The type of safety case that would have to be made is very similar. The main difference is that the spent fuel will contain all the uranium and plutonium. Therefore the total activity is about a factor of ten higher than that of the equivalent amount of high-level waste after a few thousand years. Further, the spent fuel will contain all the volatile fission product species such as iodine, I-129, which would be volatilized during reprocessing and then either discharged or separately trapped. After half a million years, the total activity will be dominated by that of the uranium decay series, and therefore will be higher than that of high-level waste from which the uranium has been removed.

The second phase of the Swedish KBS project considered the disposal of unreprocessed PWR oxide fuel (KBS 1983). This scheme relies on the durability of a strong copper canister; copper is stable under the reducing conditions typical of groundwater at depth in the Swedish bedrock and the canister should survive for at least 100 000 years. The possibility of a canister being damaged by a sudden rock movement was estimated from studies of former fracture displacements on exposed outcrops to occur once in a million years. On reasonable average assumptions, the maximum radiation dose to nearby residents was calculated to be due to I-129 and to be between 0.001 and 0.0001 mSv, occurring in about a million years. This low result is mainly due to the excellent durability expected from the thick copper canister. While the same engineering would not be applicable in all rock formations, the disposal of spent oxide fuel should not give rise to greater concentrations of radionuclides than would other forms of high-level waste because of the very low solubility of uranium and plutonium in the chemical conditions of a well engineered repository.

### Disposal in the deep sea sediments

One other location that has been considered for the disposal of high-level wastes is the sediments of the deep ocean floor, which are very old, stable strata. The function of the sediments would be to provide the 'near-field' control of the movement of radionuclides that is accomplished in land-based repositories by the engineered barriers of bentonite and cement (Francis *et al.* 1986). For this function to be efficient, the movement of pure water through the sediments should be slow and the chemistry reducing rather than oxidizing. Finally, dispersion in the oceans would delay any radionuclides that did reach the sea-floor from entering any food chain.

Two engineering concepts have been explored. In the first, high-level waste would be packed into a streamlined projectile which would be freely dropped on to the ocean-floor and end up buried in the sediments to a depth of a few tens of metres. Alternatively, a hole might be drilled into the lithified sediment or basement of the ocean-floor, waste canisters stacked within it and the hole sealed by cement grouting or other backfilling method. This is a more conventional concept which has been proposed for shallow waters (see below) but its feasibility in depths over 2 km still has to be proved.

Ten countries have taken part in a sub-sea-bed disposal programme which has been coordinated by the Sea-bed Working Group of the Nuclear Energy Agency of the OECD. The work has concentrated around the Great Meteor East area, an area of deep sediments between the Azores and the Canary Islands. Model penetrators have been dropped to about 30 m into the sediments. Measurements are being made to determine the rate of movement of pore water, and to determine the chemical conditions that pertain in the sediments at depth. It seems, however, that some years' more work will be required before the knowledge of these parameters is sufficient to enable a reliable calculation of radionuclide migration within the sediments to be made. In addition, any disposal operation under the deep-sea-floor would have to be shown to be legal within the 1972 London Dumping Convention or within the proposed UN Convention on the Law of the Sea. Recent developments within the London Dumping Convention concerning disposal of low-level waste on the sea-floor (p. 165) show what difficulties might emerge in international discussions.

### Actinide separation and destruction

Before leaving the question of high-level waste disposal, it is appropriate to mention the possibility of destroying the long-lived actinide nuclides by irradiation. It has been argued that, since the actinide elements plutonium, americium, and neptunium are potentially responsible for a significant part of the radiation dose that might be suffered in the long term, it would be advantageous to separate them and irradiate them in a fast reactor or in a beam of charged particles from an accelerator to cause them to undergo fission, replacing the long-lived activity with a mixture of nuclides of shorter half-life. While this concept seems feasible scientifically, the chemistry of the separation process required is complex and not fully developed, and any gain from the decrease in long-lived activity would have to be balanced against the increase in radiation levels in the short- to medium-term, the increased amounts of low- and intermediate-level waste, and the increase in radiological risks run by the operators (RWMAC 1985). The confidence in the ability of the engineered barriers to retain the actinide elements within a repository, which has been a feature of recent research, militates against the need for more complex methods of reducing the possible radiation doses from the actinides. A good deal of the benefit would in any case be achieved by the conventional reprocessing route, removing the plutonium along with the uranium.

### The disposal of intermediate-level wastes

Those categories of intermediate-level wastes that contain appreciable amounts of long-lived nuclides have to be disposed deep underground, as do the high-level wastes. Since the activity per unit volume is lower than that of the high-level waste by a factor of about 250, there is much less need to consider heat emission, and the packaged waste can be packed more tightly in a repository. The different waste streams have widely differing chemical and radionuclide compositions, and each must be packaged appropriately. One of the most highly active of the intermediate-level wastes is the fuel element cladding, which will contain some residual irradiated fuel, with its associated fission products and actinides, that was not removed with the bulk of the fuel during reprocessing. But the radioactive isotopes formed by the interaction of neutrons with the structural components of the fuel assemblies may be the more important nuclides—species such as Co-60, with a half-life of 5.3 years or the longer-lived nickel isotopes, Ni-63 (half-life 100 years) or Ni-59 (half-life 75 000 years).

The principles of containment using multiple barriers are the same as for high-level wastes, but the details will differ. A new feature which does not enter into a high-level waste repository is the inclusion of a variety of organic materials in some of the intermediate-level wastes. In a radiation field, these may decompose, evolving hydrogen and other gases, and possibly forming organic compounds that could enhance the solubility of uranium and plutonium. Specific tests on actual waste compositions are required to validate any proposed method of packaging. The geological conditions that have to be satisfied for a repository containing long-lived nuclides are similar to those already discussed for high-level waste. A good example is a site based on a former iron ore mine at Konrad in Germany, with a capacity for 500 000 $m^3$ of intermediate- and low-level waste. The planned repository would be at 100 m depth and be overlaid by relatively impermeable strata. The maximum estimated radiation doses are about 0.01 mSv/y, possibly arising in 300 000 years' time (Warnecke and Brennecke 1987).

## Disposal of low-level radioactive wastes

Apart from the uranium mill tailings, low-level wastes represent the largest volumes but the lowest potential hazard, less than those involved in dealing with many categories of industrial waste. Consequently more diverse methods of disposal have been adopted. Countries with large areas of dry underground mines tend to use them, while countries with areas of suitably retentive surface geology have used deep pits or trenches as repositories. The Germans plan to use the Konrad mine for LLW as well as for ILW. Artificial underground repositories are being constructed in Sweden and in Switzerland. The Swedish plan is to consign all intermediate-level waste and most of the low-level and decommissioning wastes to a repository in the bedrock some 30 m below the sea-bed off the coast at Forsmark, reached by a tunnel from the shore. In addition, up to 20 000 $m^3$ of low-level wastes will be placed in shallow burial trenches at Oskarshamm, Forsmark, and Studsvik, with strict limits both in the specific activity and the total activity to be accommodated (RWMAC 1986).

The simplest form of trench disposal has been carried out in the UK in Drigg in Cumbria. About 500 000 $m^3$ of a variety of low-level wastes have been deposited with little final packaging in trenches up to 10 m deep dug down to a level of glacial clay, angled so as to direct water that has percolated through the waste towards a stream leading to the sea. Trenches that are full are capped with a 1 m layer of earth. The site authorization conforms to the definition of low-level waste averaged over a day's disposal, but in practice the waste buried at Drigg is normally well below those limits and it has not been the practice to accept any alpha-contaminated waste other than the naturally occurring elements uranium, radium, and thorium. Radiation doses arising from the Drigg site have been assessed as being very low, but the arrangements were criticized as being insufficiently monitored and too haphazard by the Environment Committee of the House of Commons in their first report (1986). Future plans for Drigg include the compaction of waste into bales and drums, lining the trenches with concrete, capping the trenches with impermeable layers to reduce water percolation, and improving the site drainage system (BNFL 1988), a plan which is similar to that operated at the French La Manche site on the Brittany coast.

The La Manche site accepts both low- and intermediate-level waste, but only that containing short-lived species. The average content of long-lived, alpha-emitting species must be less than 0.37 GBq/t and the maximum value for one package is ten times as high, i.e. 3.7 Gbq/t, which corresponds to the UK definition of low-level waste. The wastes with high levels of beta-gamma activity are packed into concrete boxes, with drums placed between them; more drums are placed on top of form the basis of a tumulus. Backfilling material is poured over the entire pile and the artificial tumulus then covered with a thick layer of impermeable clay. The La Manche site can receive 400 000 m$^3$ of wastes and it will be full in the 1990s. A second site for trench disposal, using the same methods, has been identified about 150 km south-east of Paris after hydrogeological investigations in many areas, and procedures for the authorization of this site are well advanced. The new centre would have a capacity of 1 million m$^3$ of waste (Barthoux 1987). The French expect to retain administrative control of repository sites for 300 years after final closure, thereby minimizing any possibility of disruption caused by human intrusion until the short-lived species have decayed.

### Disposal on the deep sea bed

Another option that has been exploited for the disposal of low-level wastes, and of some that would now be categorized as intermediate-level, was to sink suitable packages on to the bed of the deep Atlantic. Disposal of highly active material is prohibited by the terms of the London Dumping Conventions, with a limit defined as 37 GBq/t for alpha-emitters, a hundred times as much for beta-gamma emitters and a million times as much for tritium. From 1978 to 1983, the sea disposal operation was carried out under an OECD/NEA multilateral consultation and surveillance mechanism at an internationally agreed site, 4 km deep, in the north-east Atlantic.

Several nations took part; the UK disposed of about 2000 t a year by this method.

The operation planned for 1983 in this country was prevented by the action of the National Union of Seamen. The Government's response was to act jointly with the TUC and appoint a committee under Professor Holliday to report on the matter. The committee found no evidence of significant harm from past disposal but recommended waiting for the results of three international reviews due in 1985 (Holliday 1984). The results of these and other studies showed that, even on very conservative and pessimistic assumptions, the maximum radiation dose that would accrue to any individual as a result of all disposals to date would be 1/50 000 of 1 mSv/year, well below the level of dose thought to be of no significance. The collective dose commitment from the sea disposal of low-level solid wastes to the entire world population over the next 10 000 years was calculated to be less than the annual dose arising from the natural radioactivity of the sea (Camplin and Hill 1986).

The Holliday Committee also recommended that a comparative study of all disposal options be carried out. Such a study was carried out and the results published by the Department of the Environment (1986). One conclusion of that study is that sea-disposal would be the preferred option for several categories of waste—namely, magnox ion-exchange resins, low-level plutonium-contaminated waste, wastes containing tritium, de-commissioning wastes and those packaged for the 1983 sea disposal consignment—with the exception of some with a high content of C-14.

Despite these favourable assessments, it seems unlikely that this option will be available in the foreseeable future. The contracting parties to the London Dumping Convention remain deeply divided on the issue. Despite the results of the technical reviews, a majority of signatories voted in 1985 for an indefinite moratorium on the disposal of any radioactive waste on the sea-bed, and it seems that any such exercise could only be carried out against considerable political opposition.

**Decommissioning wastes**

The volumes of decommissioning wastes add a significant fraction to the total volume of wastes from the nuclear fuel cycle (Table 3.8) but not to the total radioactivity. Once the fuel assemblies are removed, the remaining structure contains little long-lived activity except the metallic activation products such as Ni-63. One estimate of the decommissioning wastes from a large PWR gives a total mass of 5300 t of steel and 14 200 t of concrete. The total radioactivity is about 5 TBq, which is small compared to the total arisings during the reactor lifetime of 40 years, and the average specific activity will be less than 1 GB/t (Flowers *et al.* 1986), so most of these wastes conform to the definition of 'low-level wastes'. The total activity would decline by more than a factor of ten in a hundred years, owing to the decay of Fe-55 and Co-60, which may be an argument for delaying decommissioning, but the longer-lived nickel isotopes, Ni-59 and Ni-63 will remain. Decommissioning wastes from gas-cooled reactors will include the graphite moderators, which will contain some C-14, but the total activity should be low. The disposal of these wastes will present no new technical problems, in the sense that packaging and disposal routes will be similar to those used for the wastes arising during operation.

## Future plans for waste disposal in Britain

Until recently, the plan for intermediate- and low-level waste disposal in Britain was to establish a second site for trench burial of low-level wastes to relieve the pressure on Drigg in the early 1990s and to develop a deep site for intermediate-level wastes (White Paper, Cmnd 8607, 1982). This policy was supported by the Best Practicable Environmental Option (BPEO) study made by the Department of the Environment, in which approximate estimates were made of the costs, occupational dose, and individual and collective doses resulting from the different options of disposal for a number of categories of wastes. The individual doses calculated in this study were all extremely small, well below the level thought to be negligible, and the only regional or global collective doses of any significance arose from wastes with a high proportion of C-14. One of the conclusions of the study was that near-surface disposal is the BPEO for over 80 per cent by volume of all the waste considered. Nirex therefore started drilling operations on two coastal and two inland sites in the clay areas of eastern England in 1986, under a Special Development Order passed by Parliament, with the aim of deciding whether any one of them was suitable for the site of a low-level waste repository using an engineered trench design reminiscent of that used at the French La Manche site.

However, Nirex informed the Secretary of State for the Environment in May 1987 that the costs of site exploration and of fully engineered trench disposal had risen to the point where little savings could be foreseen compared to the alternative of burying low-level waste together with intermediate-level waste in a deep repository, considering the marginal additional costs of extending a deep site which was in any case a necessary development. This change of policy may be politically welcome since the Environment Committee of the House of Commons expressed some reservations concerning any trench disposal in their 1986 Report, on the grounds of public disquiet, although it acknowledged that the safety of fully engineered near-surface sites comes close to that of deep disposal.

So the immediate plans are to continue to use the upgraded Drigg site for LLW, perhaps introducing techniques for reducing the volumes to be disposed there, and to store other wastes until a deep depository is ready.

Nirex has been instructed to bring forward proposals for suitable deep sites for evaluation. A site would need to be some 100 acres in extent, at a depth between 200 and 1000 m; it could be under land, under the sea-bed but accessed from the cost, or under the sea-bed, accessed from an offshore structure. Areas of the UK which might provide suitable hydrogeological environments have been delineated in a recent paper (Chapman *et al.* 1986). There are many possibilities, and the major difficulty in making progress will be to find one or more which can command public support, or at least public acceptance (Roberts 1988).

Nirex circulated widely a descriptive document, *The Way Forward*, to promote public discussion of the issues involved, and an analysis of the replies from local authorities, public bodies, and members of the public has been published (Environmental Risk Assessment Unit, 1988). It was announced in March 1989, that Nirex would first seek to evaluate two sites in areas where they had found a measure of public support—Sellafield, in Cumbria, and Dounreay, in Caithness. (Nirex 1989*a*). Nirex has also published a preliminary environmental and radiological assessment of these two sites (1989*b*); if neither is found to be suitable, then further sites will have to be explored.

## Global impact of a nuclear power programme

The discussion of the individual parts of the nuclear fuel cycle in the preceding sections of this Chapter shows that the radiological impact of the nuclear fuel cycle under the conditions of normal operation will be low, as shown by calculations of collective, committed radiation dose made on a world-wide basis by UNSCEAR (1988). The annual public collective dose from the present nuclear programme of 190 GW per year is assessed to be 760 man Sv, about 0.01% of that from natural background sources. In the distant future, doses will be due mainly to the uranium mill tailings, with a smaller contribution from long-lived radionuclides such as I-129. These doses may be further reduced by technological improvements and by the use of breeder reactors to reduce the dependence on fresh uranium supplies. For comparison, as we have already seen, the annual collective dose to the general population of the UK from all the operations of the nuclear industry was 30 man Sv in 1987 (Hughes 1988), while workers in the nuclear industry accumulated a collective dose of 92 man Sv. In the same year, the collective dose to the UK population from flying in aeroplanes was 500 man Sv, and miners in coal and other mines accumulated 124 man Sv.

A comparison of the radiological impact of all the various steps in the nuclear fuel cycle leads to the conclusion that the final disposal of solid wastes is the safest step in the whole cycle. Any possibility of rapid dispersion into the environment is removed once wastes are incorporated into durable solid forms, and the work already done on the packaging of these wastes and the proving of disposal options shows that stringent safety conditions can be met. There is no credible scenario that could lead to an environmental disaster. This does not seem to be the popular perception and it is most regrettable that the subject of waste disposal has aroused so much unnecessary alarm.

The critical question to be considered, therefore, in assessing the environmental impact of a nuclear power programme, is the safety of nuclear reactors and of other plants; serious environmental effects only arise as a consequence of major accidents. The attainment of acceptable standards in nuclear energy depends on a careful adherence to high standards in design, in construction and in operation; nuclear energy is no different in this respect from other high-technology industries such as petrochemicals or air transport. As in these other industries, safety must be seen as an international problem requiring internationally-agreed standards and proper means of inspection and enforcement. The management concerned and the regulatory authorities must command public respect and support. The size of the nuclear energy programme in the future depends more on that consideration than on any technical question and this matter will be discussed further in the final chapter.

# References

ACTRAM (1988a). *The UK Regulations on the Transport of Radioactive Materials.* HMSO, London. ACTRAM (1988b) *The Transport of Civil Plutonium by Air.* HMSO, London.

Allardice, R.H., Harris, D.W., and Mills, H.L. (1983). *Nuclear Power Technology*, **2**, p. 209, Oxford University Press.

Anthony, R.D. (1986). Safety and licensing; the British perspective. *Nuclear Energy*, **25**, p. 29.

Ashworth, F.P.O. and Western, D.J. (1987). Sizewell B degraded core analysis. *Nuclear Energy*, **26**, p. 233.

Avery, D.G. (1984). Liquid discharges from the Sellafield site, evidence to the Sizewell Inquiry. BNFL/P/1 (ADD 12).

Barthoux, A. and Vigreux, B. (1987). Radioactive waste disposal in France. *Nuclear Tech. International*, p. 193.

BEIR (1980). The effects on populations of exposure to low levels of ionizing radiation. *Report of the National Research Committee on the Biological Effects of Ionizing Radiation (BEIR)*. National Academy Press, New York.

Beral, V. *et al.* (1985). Mortality of employees in the United Kingdom Atomic Energy Authority, 1946–1979. *British Medical Journal*, **291**, p. 440.

Bridge, M.J., Gallie, R.R., Hewitt, P.V. and Perceival, K. (1982). Concluding experiments on the Windscale AGR. *Nuclear Energy*, **21**, p. 29.

British Nuclear Fuels plc. (1988). Evidence to the Sub-Committee of the Select Committee on the European Communities of the House of Lords.

Broadley, D. (1986). Commercial demonstration fast reactor safety. *Nuclear Energy*, **25**, p. 101.

Bromley, J. (1986). A Comparison of the Hazards of Mining Coal and Uranium. Uranium Institute, Eleventh Annual Symposium, London.

Brown, M.L. (1986). Safety Aspects of Transport. Precognition to Public Inquiry, European Demonstration Reprocessing Plant.

Bush, D. (1982). Radiological protection associated with normal operation of PWRs. *Nuclear Energy*, **21**, p. 395.

Camplin, W.C. and Hill, M.D. (1986). Sea dumping of solid radioactive waste, in *Radioactive Waste Management and the Nuclear Fuel Cycle*, **7**, p. 233.

Chapman, N.A., McEwen, T.J. and Beale, H. (1986). *Geological environments for deep disposal of intermediate level wastes in the UK*. IAEA-SM-289/37. IAEA, Vienna.

Clarke, R.H. (1988). *Radiological Protection Bulletin* No. 95, p. 7.

Cogne, F. and Justin, F. (1985). Safety of the Creys-Malville Plant. *Nuclear Europe*, **11**, p. 23.

COMARE First Report (1986). The implications of the new data on the releases from Sellafield in the 1950s for the conclusions of the Report on the Investigation of the Possible Increased Incidence of Cancer in West Cumbria. HMSO, London.

COMARE Second Report (1988). Investigation of the possible increased incidence of leukaemia in young people near Dounreay Nuclear Establishment, Caithness, Scotland.

Cook-Mozaffari, P.J. *et al.* (1987). Cancer incidence and mortality in the vicinity of nuclear installations in England and Wales, 1959–1980. *Studies of Medical and Population Subjects*, No. 51. OPCS.

Corbet, A.D.W. *et al.* (1986) Control and reduction of liquid radioactive discharges from the Sellafield reprocessing site. ENC Conference, Geneva. **4**, 525. European Nuclear Energy Society, Berne.

Dale, G.C. (ed.) (1982). Safety of the AGR. CEGB and SSEB, London and Glasgow.

Darby, S.C. *et al.* (1985). Mortality of employees in the United Kingdom Atomic Energy Authority, 1946–1979. *British Medical Journal*, **291**, p. 272.

Day, D.H., Hughes, A.E., Leake, J.W., Marples, J.A.C., Marsh, G.P., and Rae, J. (1985). The management of radioactive wastes. *Rep. Prog. Phys.*, **48**, p. 101.

Dean, R.A. and Simon, W.A. (1987). Modular high temperature gas cooled reactor. *Nuclear Tech. International*, **p. 53.**

Department of the Environment (1984). *Disposal facilities on land for low- and intermediate-level radioactive wastes*. HMSO, London.

Department of the Environment (1986). *Assessment of best practicable environmental options for the management of low- and intermediate-level radioactive wastes*. HMSO, London.

Department of the Environment (1987). Report DOE/RW/88.036. UK Natural Analogue Co-ordinating Group.

Duffy, L.P., Kinther, E.E., Fillnow, R.H., and Fisch, J.W. (1986) The Three Mile Island accident and recovery, *Nuclear Energy*, **25**, p. 199.

Edwards, A.R. (1985). Approaches to containment. *Nuclear Energy*, **24**, p. 241.

Elsden, A.D. and Heafield, W. (1985). In *Radioactive Waste Management*, p. 139 BNES, London.

Environmental Risk Assessment Unit, *Responses to The Way Forward*, (1988), University of East Anglia, Norwich.

Eve, A.S. (1939). *Rutherford*. Cambridge University Press.

Flowers, R.H. and Chapman, N.A. (1986). *Phil. Trans. Roy. Soc. Lond.*, **A319**, p. 83.

Flowers, R.H., Roberts, L.E.J., and Tymons, B.J. (1986). *Phil. Trans. Roy. Soc., Lond.*, **A319**, p. 5.

Forman, D. *et al.* (1987). Cancer near nuclear installations. *Nature*, **329**, p. 499.

Francis, T.J.G., Searle, R.C., and Wilson, T.R.S. (1986). *Phil. Trans. Roy. Soc. Lond.*, **A319**, p. 157.

Frigerio, N.A. *et al.* (1973). *Carcinogenic hazard from low-level, low-rate radiation*. Argonne Radiological Impact Programme (ARIP) Part 1, ANL/ES-26 (Pt 1), Argonne National Laboratory, Ill, USA.

Gittus, J.H. *et al.* (1988). *The Chernobyl accident and its consequences*. UKAEA Report NOR 4200, 2nd ed., HMSO London.

Godfrey, D.G., Davies, I.L. and MacKenzie, R.D. (1985). A review of construction techniques available for surface and underground waste repositories. *Rad. Waste Management*, BNES, London, p. 239.

Hart, J.D., Blythe, R.A., Milne, I., and Shears, M. (1985). In *Seminar on the resistance to impact of spent magnox fuel transport flasks*, p. 79. Institute of Mechanical Engineers, London.

Hennies, H.H. (1986). LWR safety technology in Europe. *ENC '86 Transactions* p. 167. European Nuclear Society, Berne.

Hodgkinson, D. P., Lever, D. A., and Rae, J. (1983). *Progress in Nuclear Energy*, **11**, p. 183.

Holliday, F.G.T. (1984). Report of the Independent Review of the disposal of radioactive waste in the north-east Atlantic. HMSO.

House of Commons (1986). First Report from the Environment Committee, radioactive waste, HMSO, London.

Howarth, J.M. and Eggleton, A.E.J. (1987). Studies of environmental radioactivity in Cumbria. AERE-R11733. AERE, Harwell, Oxfordshire.

HSE (1978). *Transport of plutonium nitrate solution between Dounreay and Windscale*. Report NII/R/39/78. HSE.

HSE (1986). *Safety audit of BNFL, Sellafield*. HMSO, London.

HSE (1987). *The tolerability of risk from nuclear power stations*. HMSO, London.

HSE (1987). *Bradwell Nuclear Power Station*. HMSO, London.

Hughes, A.E., Marples, J.A.C., and Stoneham, A.M. (1983). *Nuclear Technology*, **61**, p. 496.

Hughes, J.S. and Roberts, G.C. (1984). *The radiation exposure of the UK population — 1984 review*. NRPB-R173, HMSO, London.

Hughes, J.S. (1988). *The radiation exposure of the UK population — 1988 review*. NRPB-R227, HMSO London.

Hunt, G.J. (1987). Radioactivity in surface and coastal waters of the British Isles, 1986. M.A.F.F. Report No. 18, Lowestoft; Hunt, G.J. (1988). *Radioactivity in surface and coastal waters of the British Isles.* M.A.F.F. Report No. 19, Lowestoft.

IAEA (1981). Current practices and options for the confinement of uranium mill tailings. *Tech. Rep. Series*, No. 209. IAEA, Vienna.

IAEA (1986). General conference paper GC (SPL.1/3), 24 September.

ICRP (1975). Report of the Task Group — Reference Man. *ICRP*, **23**, Pergamon Press.

ICRP (1976). Recommendations of the International Commission on radiological protection. ICRP 26, *Annals of the ICRP*, **1**(3), p. 1.

ICRP (1984). Principles of limiting exposure of the public to natural sources of radiation. *Annals of the ICRP*, **14**(1), p. 1.

ICRP (1987). Publication 50. Lung cancer risk from indoor exposures to radon daughters. A Report of a Task Group of the ICRP. *Annals of the ICRP*, **17**, No. 1.

J.C. Consultancy Ltd. (1986). *Risk assessment for hazardous installations.* Pergamon Press, Oxford.

KBS (1978). (Kärn-Bränsle-Säkarket). Handling of spent nuclear fuel and fuel storage of vitrified high level waste. SKBF/KBS Stockholm.

KBS-3 (1983) Final Storage of Spent Nuclear Fuel. SKBF/KBS Stockholm.

Kemeny, J.G. (1979). Report of the President's Commission on the accident at TMI. Washington D.C.

Kinlen, L.J. (1988). *Lancet*, p. 1323. 10 December 1988.

Kinlen, L.J. (1989). In *medical response to effects of ionizing radiation* (ed. W. Crosbie and J. Gittus). Elsevier.

Kühn, K. (1986). Field experiments in salt formations. *Phil. Trans. Roy. Soc. Lond.*, **A319**, p. 157.

Layfield, Sir Frank (1987). Sizewell B Public Inquiry. Department of Energy Report, HMSO London.

Leclerq, J. (1986). *The nuclear age,* Hachette.

Lewis, H.W. (1978). Risk-Assessment Review Group Report to the USNRC Report NUREG-CR-0400. Nuclear Regulatory Commission, Washington D.C.

Luckey, T.D. (1980). *Hormesis with ionizing radiation* C.R.C. Press.

Marples, D.R. (1986). *Chernobyl and nuclear power in the USSR* Macmillan, London.

Marsham, T. (1985). *Nuclear Energy*, **24**, p. 155.

Mendel, J.E. (1986). Fixing high level wastes in glasses. *Phil. Trans. Roy. Soc. Lond.*, **A319**, p. 49.

Morse, J.A. and Wilkinson, A. (1985). In *The resistance to impact of spent magnox fuel transport flasks*. p. 65, Institute of Mechanical Engineers, 1 May 1985.

Muller, H.J. (1927). The problem of genetic modification. 5th International Genetics Congress of Berlin. *Z. Ind. Abst. Vererblarke*, **1** (Suppl. 1), p. 234.

Mummery, P. (1985). Control of liquid discharges from Sellafield. *Nuclear Europe*, **19**.

Mummery, P. and Anderson, A. R. (1987). In *ALARA, Principles, practice and consequences*, p. 104. Adam Hilger, Bristol.

Nambi, K.S.V. and Soman, S.D. (1987). *Health Physics*, **28**(5), p. 653.

Nirex (1987) *The way forward*. UK Nirex Ltd.

Nirex (1989*a*) *Going forward* UK Nirex Ltd.

Nirex (1989*b*) Deep Repository Project. Nirex Report No. 71.

NRPB (1985). Small radiation doses to members of the public. ASP7.

NRPB (1986). Living with radiation.

NRPB (1987*a*). Statement from the 1987 Meeting of the International Commission on Radiological Protection. Supplement to *Radiological Protection Bulletin* No. 86.

NRPB (1987*b*). Interim guidance on the implications of recent revisions of risk estimates and the ICRP 1987 Como Statement. Report GS9.

NRPB (1987*c*). Exposure to radon daughters in dwellings. Report ASP 10.

NRPB (1988). Health effects models developed from the 1988 UNSCEAR Report Report R226.

Okrent, D. (1987). The safety goals of the US Nuclear Regulatory Commission. *Science*, **236**, p. 296.

Otake, M. and Schull, W.J. (1984). *In utero* exposure to A-bomb radiation and mental retardation: a reassessment. *Brit. J. Radiol.* **57**, p. 409.

Parker, F.L., Kasperson, R.E., Anderson, T.L., and Parker, S.A. (1987). Technical and socio-political issues in radioactive waste disposal. The Beijer Institute, Stockholm.

Parker, M. Justice. (1978). The Windscale Inquiry Report. HMSO London.

Pochin, E. (1983). *Nuclear radiation: risks and benefits*. Clarendon Press, Oxford.

Pochin, E. (1987). Summarization meeting of the work of the Chinese High Background Radiation Research Group. NRPB Bulletin **102** p. 11.

Posner, E. (1970). Reception of Röntgen's discovery in Britain and USA. *Brit. Med. J.*, **4**, p. 357.

Rae, J. and Black, J.H. (1986). Modelling of radionuclide migration. *Phil. Trans. Roy. Soc., Lond.*, **A319**, p. 83.

Rasmussen, N. (1975). *Reactor Safety Study: an Assessment of Accident Risk in U.S. Commercial Power Plants*. Report WASH-1400. Nuclear Regulatory Commission, Washington, D.C.

Ringwood, A.E. and Kelly, P.M. (1986). Immobilization of high level waste in ceramic wasteforms. *Phil. Trans. Roy. Soc., Lond.*, **A319**, p. 63.

Roberts, L.E.J. (1979). Radioactive waste disposal — policy and perspectives. *Nuclear Energy*, **18**, p. 85.

Roberts, L.E.J. (1984). *Nuclear power and public responsibility*. Cambridge University Press.

Roberts, L.E.J. (1988). Radwaste, spectre, or symbol? *Roy. Inst. Proc.* p. 259.

Roentgen Society (1915). *Journal of the Roentgen Society*, p. 113.

Rose, K.S.B. (1982). Review of health studies at Kerala. *Nuclear Energy*, **21**(6), p. 399.

RCEP (1976) (Royal Commission on Environmental Pollution). Sixth Report, Cmnd 6618. HMSO, London.

RWMAC (1984) (Radioactive Waste Management Advisory Committee). Fifth Annual Report, HMSO, London.

RWMAC (1985) (Radioactive Waste Management Advisory Committee). Sixth Annual Report, p. 45. HMSO, London.

RWMAC (1986) (Radioactive Waste Management Advisory Committee). Seventh Annual Report, p. 51. HMSO, London.

Salmon, A. (1984). Transportation of spent nuclear fuel. *Nuclear Energy*, **23**, p. 237.

Saunders, P.A.H. and Wade, B.O. (1983). Radiation and its control, in *Nuclear Power Technology*, **3**. Clarendon Press, Oxford.

Shabad, T. (1986). Geographic aspects of the Chernobyl nuclear accident. *Soviet Geography*, **XXVII**, p. 504.

Shaw, K.B., Mairs, J.H., Gelder, R., and Hughes, J.S. (1984). NRPB REPORT R155, HMSO, London.

Shimizu, Y. *et al.* (1987). Radiation effects research foundation report RERF TR. pp. 12–87.

Smith, P.G. and Douglas, A.J. (1986). Mortality of workers at the Sellafield Plant of British Nuclear Fuels. *Brit. Med. J.*, **293**, p. 845.

Spaargaren, F.A. (1988). Searching for uranium. *The Nuclear Engineer*, **29**, p. 127.

Stone, R.S. (1952). The concept of a maximum permissible exposure. *Radiology*, **58**, p. 638.

Tolley, H.D. *et al.* (1983). A further update of the analysis of mortality of workers in a nuclear facility. *Radiation Research*, **95**, p. 211.

UNSCEAR (1982). *Ionizing radiation: sources and biological effects.* United Nations Scientific Committee on the Effects of Atomic Radiation, United Nations, New York.

UNSCEAR (1986). *Genetic and somatic effects of ionizing radiation.* United Nations Scientific Committee on the Effects of Atomic Radiation. United Nations, New York.

UNSCEAR (1988). *Sources, effects and risks of ionizing radiation.* United Nations Scientific Committee on the Effects of Atomic Radiation, New York.

Uranium Institute (1984). *Principles of uranium mill tailings isolation and containment.* Uranium Institute, London.

Warnecke, E. and Brennecke, P. (1987). *Kerntechnik*, **51**, p. 87.

Weinberg, A. and I. Spiewak. (1985). *Ann. Rev. Energy*, **10**, p. 431.

White Paper Cmnd 8607 (1982), *Radioactive Waste Management*, HMSO, London.

Williams, T.I. (ed.) (1978). *A History of Technology*. Clarendon Press, Oxford.

# 4

# PROSPECTS FOR THE FUTURE

The large scale and high capital cost of all electricity generating systems results in considerable inertia; it is unlikely that there will be any sudden or dramatic change in the environmental impact per unit of generation, and the total global impact will depend for some time on the level of demand and the proportion of power stations that burn fossil fuels. Large improvements in the overall efficiency of electrical generating stations are a distant prospect. It is probable that overall demand will continue to increase given a rising world population, particularly in countries anxious to improve both their standard of living and the level of industrialization. Increasing capacity will be offset to some extent by further improvements in the control of emissions, especially from coal-fired stations. In the short to medium term, economic pressures and a buoyant gas supply may lead to the introduction of comparatively small generating stations burning gas which can be sited close to centres of population more easily than the very large coal-burning or nuclear plants. The environmental impact of such stations would be low, and a switch to burning gas rather than coal or oil would be advantageous from the point of view of lowering carbon dioxide emissions for a given output of energy (p. 47). However, the increasing pressure on energy supplies that will develop in the next century as oil and, later, gas supplies decline will force increasing emphasis on measures of conservation, on the exploitation of the renewable sources of energy and on any new developments that hold out the promise of tapping large new resources. One way of increasing the overall efficiency of fuel use is to arrange to use the waste heat from power generation to meet a local need for heat in combined heat and power schemes and we shall look briefly at the environmental impact of such schemes in the next section.

The 'renewable' sources of energy are direct or indirect means of exploiting the heat from the sun, with the exception of tidal energy, which depends on the gravitational pull of the moon, and of geothermal energy, which is dependent on the heat stored in the earth resulting from radioactive decay over geological times. The electricity-providing renewable technologies include tidal power, wind, and wave power, photovoltaics and hydroelectricity. In these cases there is no need for fuel supply and no effluent; the environmental impacts arise because of the nature of the sites required and of the comparatively low energy density of the energy supply, leading to large structures and high requirements for land. Burning wood or agricultural refuse is unlikely to be used for electricity generation, except for small-scale local schemes for combined heat and power. The use of geothermal steam might involve effluent problems, which are discussed below.

Pooley (1987) has summarized the economic and commercial prospects for renewable energy resources in the UK. The only electricity-producing technology that is economically attractive at present is conventional hydropower. Small-scale hydropower, on-shore wind, tidal power, and geothermal power from hot, dry rocks show considerable promise but the economics and possible timescale of market penetration remain to be demonstrated; all are worth developing. On the other hand, off-shore wind schemes, wave power, and the photovoltaic generation of electricity are rated as 'long shots', where it seems unlikely that the technology can be sufficiently improved or that fuel prices will rise sufficiently to make them economically attractive in the UK in the foreseeable future except for small, local schemes.

Photovoltaic cells are used in small, specialist applications such as communication satellites and they are clean, non-polluting, and easy to maintain. However, the costs of electricity produced from them in larger-scale applications is more than ten times the current production cost for electricity, and the costs of mounting the large units required and of power conditioning equipment would probably rule them out for anything but specialist applications in Britain unless the efficiency of solar cells can be increased considerably and hence the area required reduced. Even in Australia, where there is more need for power in remote areas and more sunshine, solar cells are unlikely to displace diesel/battery systems; the Australian Department of Resources and Energy (1985) has estimated that photovoltaic applications could only displace 0.1 per cent of Australia's current annual oil consumption.

Similarly, wave energy converters are unlikely to prove economic except for relatively small-scale applications, probably close inshore. Wave energy converters built to withstand the worst sea conditions well offshore on the West coasts of Britain would necessarily be very large, strong structures, with an electrical output per unit weight of structural material about 5 per cent of that from a nuclear power station or land-based wind turbine. They are therefore uncompetitive and might also present several environmental hazards, including dangers to shipping and difficult maintenance problems. Further, the possible effects on the local wave climate and consequential ecological changes, especially in fisheries, would have to be evaluated carefully.

We shall therefore concentrate in the rest of this Chapter on those renewable resources that seem likely to make a significant contribution to electricity generation in the UK in the next forty years, given successful development. Schemes that depend on large arrays of solar collectors to raise steam are unlikely to succeed in the British climate. Table 4.1

**Table 4.1**

Contribution of renewable technologies in the UK
(Department of Energy 1988)

|  | Technical potential TWh/y | Estimated contribution by 2025, TWh/y |
|---|---|---|
| Wind power, On-shore | 45 | 0–30 |
| Off-shore | 140 | ? |
| Tidal power | 54 | 0–28 |
| Geothermal hot dry rocks | 210 | 0–10 |
| Wave power | 50 | 0–0.2 |
| Small-scale hydropower | 2 | 0.3–0.7 |

reproduces a summary of the technical potential and estimated contribution to UK electricity generation by the year 2025 of the more promising schemes recently published by the Department of Energy (1988). The technical potential is a calculation of the total resource available in the UK, applying only very broad economic and environmental constraints. The estimated contributions take into account the state of the technical programme, the market likely to be available, and the extent of penetration of that market by 2025. The range of estimated contributions may be compared with the present generation of electricity in the UK, which is about 250 TWh/y. Both wind power and tidal power are already exploited elsewhere, and what remains to be determined is the economy and environmental impact of large-scale schemes. Geothermal power generated by raising steam from hot, dry rocks at considerable depths is at a much earlier stage of development, but is thought to be worth pursuing · because of its promise and the potential size of the resource.

Another technology that exploits a new source which may come to commercial exploitation in the next thirty to fifty years is nuclear fusion, which would depend only on supplies of hydrogen and of lithium. The possible environmental effects of fusion reactors as compared to those of present-day nuclear reactors based on the fission of uranium are discussed in the penultimate section.

## Combined heat and power

The generation of electricity by any of the conventional methods is necessarily inefficient in that about two-thirds of the primary input of heat energy is rejected, much of it as water at a temperature of 30–35°C, which is too low for economic use except for some marginal activities such as heating greenhouses or fish farms. A higher overall efficiency of fuel can be attained by arranging to reject the heat at higher temperatures if there is a convenient local use for it. The efficiency of electricity production will thereby be lowered, and the hot water will not be produced as economically as by the best dedicated boiler, but the costs of the electricity produced and of the hot water taken together can be less than the cost of buying the electricity from the grid and installing a separate supply of hot water.

Medium-scale combined heat and power (CHP) schemes with electrical outputs of several MW have been installed in industries such as iron and steel, paper making, oil refining, and bulk chemicals since the 1950s. In 1982, about 6 per cent of the total UK generation capacity was provided by industrial private generation (Forrest et al. 1985). More recently, interest has grown in CHP generations of smaller capacity, up to about 200 KW electrical output, for use in individual buildings such as hotels, hospitals, or university residences.

These 'micro-CHP' units are based on modified automotive or industrial engines, using natural gas, biogas from sewage works or liquefied gas (propane) or sometimes diesel fuel. Gas is a cleaner fuel to burn than either heavy oils or coal, and the environmental impact of the exhaust gases will be less than that from cars using petrol or diesel fuel, though there is a possibility of carbon monoxide being produced along with carbon dioxide. The only other nuisance that has to be guarded against is excessive noise. There are 120 sites with micro-CHP units installed in the UK, and the use of this technology is likely to increase, making some small contribution to energy conservation.

More substantial CHP systems have been installed to provide district heating to areas of towns and cities in several countries such as Sweden, Denmark, Germany, and Holland. The practicality and economics of introducing comparatively large-scale schemes in the UK was studied by the Combined Heat and Power Group set up by the Department of Energy (1979). This group examined a possible future for CHP systems in large cities, looking forward to the time when oil and gas will not be available for heating, and the alternatives for satisfying the demand for heat would be Substitute Natural Gas (SNG) produced from coal, or electrical heating or heat-only district heating schemes relying on coal. Given that situation, the group estimated the CHP district heating schemes could be found to be economically justified in about 30% of the domestic, commercial, and institutional heat markets, leading to an approximate saving in primary energy requirement of from 5 to 30 million tons of coal, or the equivalent, which would amount to 2 to 9% of the total demand for energy. Penetration of the market will be slow because of economic competition with natural gas, and because of the high capital costs of installing district heating mains in existing towns and cities. In 1985, the Government announced that grants of up to £250 000 each would be offered to Belfast, Edinburgh, and Leicester to prepare plans for CHP development, and there is interest in other cities also. Schemes in other countries tend to be associated with high density housing developments which are not so common here.

It seems that CHP could offer worthwhile energy savings in the UK in the future, though the actual savings to be achieved are uncertain and would depend on the competition with other technologies: developments in heat pumps, for example, might transform the situation. The environmental effects are difficult to estimate. Fewer central power stations would be required, with consequent environmental benefits, but to set against these gains would be the considerable initial disturbance caused by the introduction of district heating schemes close to centres of large populations, for it seems that CHP would only be economic in areas where the heat load per unit area is high. The size of stations considered for large-scale schemes have an electrical output of 200–660 MW and it is usually assumed they would have to be sited within about 15 km of the main heat load, though the costs of transmitting hot water are not very high and longer distances may be possible. Stations of this size burning coal would have to be carefully sited and designed to avoid the environmental detriment associated with relatively small generating stations in urban areas which was noted in the early years of the industry (Chapter 1). A 600 MW(e) station burns about 1 500 000 tons of coal a year and generates perhaps 20 000 tons of ash. Then the flue gas cleaning equipment or coal burning technology would have to be at least as efficient as those on large power stations or the local atmospheric conditions would deteriorate. CHP schemes burning gas would have a much lower environmental impact, and the gain in efficiency over electricity generation alone would offset the disadvantages of using a premium fuel, gas, in power stations. However, careful note would have to be taken of the future supply position of gas.

The local environmental effects due to transport and atmospheric effluents would be the least if the source of heat were nuclear. Unless transporting heat for long distances proved to be economic, two conditions would have to be satisfied. First, very stringent conditions and guarantees on safety would have to be given before reactors could be sited close to centres of population and secondly, economic designs of reactors in smaller sizes than are current would have to be developed. So many of the costs associated with nuclear reactors are relatively insensitive to size that the industry has tended to push towards larger sizes to achieve some economies of scale. Nevertheless, suggestions for suitable designs of small reactors have been made, with extremely durable fuel of long life and strong containment. These reactors are based either on research reactors of low power or on designs of reactors originally intended for long periods of operation in remote areas. For example, there is a suggestion that the Canadian SLOWPOKE reactor, built in various places as a source of neutrons, could be developed from 20 KW to 2 MW(th) output and used as a source of heat in large buildings. The French have developed two concepts, the CAS reactor capable of combined heat and power generation, suitable for powers to 300–1000 MW (thermal), a prototype of which is running at a research site in Cadarache, and the THERMOS reactor, designed for supplying heat only, in amounts from 100 to 200 MW. THERMOS is a low-pressure water moderated reactor with an exit water temperature of 140°C, and a highly retentive, tough fuel (Huest 1983). A long period of reliable operation would be required before any such designs could be used with confidence.

The prospect is, then, that the use of small-scale CHP units will continue to expand in the UK, and there is an incentive to develop large-scale schemes in suitable locations. There should be no bar to CHP on environmental grounds so long as the stations are carefully sited and standards of effluent control are strict. However, the rate at which large CHP systems can be constructed will depend on the introduction of district heating schemes and this will necessarily be slow, at least as long as gas prices are low.

## Hydropower

Large-scale hydroelectric schemes now account for about 20 per cent of the world's generation of electricity; 1300 MW of capacity is installed in the UK, nearly all in the North of Scotland. These schemes can have both environmental advantages and disadvantages. The building of dams and the creation of artificial lakes can aid flood control and flow regulation and create a local environment for recreation and water sports and thereby encourage tourism. But the initial impact of such a scheme can be very severe, with the flooding of large areas that may lead to the abandonment of farms or even villages, with the consequent displacement of populations. There can be negative effects also during operation, such as changes in water quality and in the ecology of rivers, with occasional adverse effects on fish.

The failure of dams can cause environmental disasters, with large loss of life and of productive land. The filling of the Vaiont Dam reservoir in Italy led to a landslip in 1963 that resulted in 2000 deaths. Coppola and Hall (1981) list 13 dam failures in the USA in the period 1874 to 1977, with the loss of over 3500 lives, while Okrent (1981) claims that 100 large dams in the USA have failed since 1930, and considers that about 40 per cent of the 50 000 public and private dams in the USA 'present a significant or large hazard potential to downstream life, if they should fail'. Apparently there has not been a formal requirement for probabilistic analyses of the safety of these dams, though there are current efforts to try to use quantitative risk acceptance criteria.

Such dangers do not attend the operation of small-scale hydroelectric schemes and these may be able to make significant local contributions in many parts of the world. Their introduction may be limited by other environmental issues, however, since the best sites are often in areas of natural beauty and the need for roads, construction works, and transmission lines may cause concern. The total technical potential for small schemes in the UK may amount to 2 per cent of national electricity consumption, though this estimate takes no account of local difficulties or of cost (Pooley 1987), and the DoE estimated contribution is 0.3 to 0.7 TWh/y (Table 4.1).

## Tidal power

Tidal power is a special case of hydroelectric generation, involving the building of an artificial barrier across an estuary with a high tidal range, with release of water through turbines. A 240 MW station has been operating at La Rance in France for over twenty years. There are several possible sites in the UK, notably the Severn Estuary, Morecombe Bay, the Solway Firth and the Wash; the most intensively studied to date has been the Severn Barrage proposal. A barrage could be built across the Severn in several places and the cost, the time taken to build it, and the environmental consequences will depend on the final choice of site. A total of 23 TWh/y might be generated at a cost of £0.05/Kwh or less from the more promising sites, which would amount to 8.5 per cent of the UK's present consumption of electricity.

Tidal barrages would be huge civil engineering projects that would take 10–15 years to complete and which would have a profound effect on the local environment and the local economy both during the construction period and thereafter. Once built, a barrage might have a working life of at least 120 years, with the plant and equipment being replaced every 40 years (Glover 1981). The effects on shipping movements, the impact on agricultural land and the possible need for new treatment plants for industrial wastes and domestic sewage would have to be evaluated before any barrage was built. The possible effects on the local ecology are likely to be complex. Changing water levels will alter the areas of mudflats at low tides, and new regimes of water flow would alter the turbidity, salinity, and oxygen levels of local waters, with consequent effects on the marine flora and plankton concentrations and hence on the fish and bird populations.

Following the publication of the Report from the Severn Tidal Power Group in 1986, the Severn Barrage Development Project issued a programme of studies on the further development of models of the estuary, allowing more accurate predictions to be made of the tide levels, land drainage requirements, sediment deposition, effluent dispersal, and water quality and ecosystem changes. The capital costs of the barrage currently considered, which would be 15 km long, would be of the order of £5000 m, and the financial and administrative problems to be tackled, given the multiplicity of interests concerned, would be as least as difficult as the environmental questions.

## Wind power

Wind generators are the most promising of the 'renewable' technologies for electricity generation for application in the near future in the UK. They are already used in other parts of the world, such as Denmark and the USA, and the British wind climate is quite favourable, so the total resource that might be exploited is large. The ultimate economics of wind generation remain to be proved as a result of the operation of the many prototype experiments now in progress but it seems likely that the costs of the electricity produced will be within the range of current costs if the wind generators are sited on land, though the costs and the technology are less certain for generators sited offshore. Wind power from sites on land could be contributing up to 10 per cent of current electricity generation, 27 Twh/y, by 2030, while Pooley (1987) gives a range of 0–10 Twh/y by 2010. The efficiency of energy collection by wind generators is better than that of wave energy converters, with a useful output per unit weight of the structure perhaps ten times higher and of the same order as for nuclear power stations; most of the structure can be built by conventional civil engineering techniques. Further, individual machines are quite small, with an output from 1–3 MW, so that modular schemes of different sizes can be developed and fitted into different sites. This flexibility is attractive compared to the huge capital investment and long-term planning and commitment typical of tidal power schemes. However, the wind climate is variable so that spare capacity based on conventional technology would have to be available to meet the demand on still days. The contribution from wind energy will therefore be to reduce our dependence on other fuel supplies, and to reduce the pollution caused by burning fossil fuel, rather than to reduce the need for generation capacity.

Small wind turbines of up to 100 KW output are used now in isolated sites, but the total contribution to national supplies is small. Significant exploitation of wind energy would require the construction of numbers of large machines, of 1 MW average output or more. Several machines with a rated output of 2–4 MW have been built in other countries (ETSU 1985). These are very large machines, with rotor blades of between 60 and 100 m in diameter, mounted on steel or concrete towers. The installation of a generating capacity of 1 GW, equivalent to a large nuclear or coal-burning station, would require the construction of about 330 machines of this size. If they were spaced about 1 km apart, they would occupy an area of 330 km$^2$, that is, a plot 18 km × 18 km or an equivalent rectangle, and the best sites would be those with the best wind climate, which tend to be upland sites.

Land disturbance would be limited to the time of construction, since the land could be used for normal purposes such as agriculture right up to the base of operating generators. The main environmental impacts would be visual intrusion, electromagnetic interference and noise. Evidence to date suggests that the hazard presented to birds will be low. But wind generators would have to be sited so as not to interfere with transmissions from TV relay stations and from microwave communication links. Public acceptability will depend on reactions to the visual appearance and to the noise, which can arise in three ways: machine noise, aerodynamic noise, and infrasound from blade/tower interactions. It is not yet clear how close generators could be sited to domestic buildings. The safety aspects should be acceptable; there is some possibility of blade fracture with the resulting missiles being thrown a distance of more than 300 m but the probability of such accidents is considered to be low (ETSU 1985).

## Geothermal energy

The exploitation of hot water aquifers is established as a heat source in several European countries but the test drillings that have been carried out to date in the UK have been disappointing; few substantial resources have been found close to a large market for heat. In any case, any contribution from hot water aquifers in the UK would only be of second order importance for electricity generation, in that it might reduce the market for the provision of heating; the temperatures at which water is recovered is too low for electricity generation.

A more promising development is to reach higher water temperatures by pumping water to greater depths, up to 6 km, where the geothermal temperature gradients are high, the so-called 'hot dry rock' technique. Two boreholes are drilled down to rocks at a useful temperature and means found to fracture the rocks at that depth so as to form a large surface for the transfer of heat from rock to fluid. Water is then pumped down one hole, flows through the fractured zone and is recovered from the other hole. The recovered water would have to be at temperatures about 170°C for electricity generation but lower temperatures might be acceptable for combined heat and power schemes or for heat-only applications if the station is situated near enough to a large heat load to make such applications viable; however, district heating schemes would have to be developed, and these will only be introduced slowly (see p. 177). Electricity generating stations can be sited more flexibly, for example in areas such as Cornwall where the geothermal gradients are high.

The work on 'hot dry rocks' is only at an experimental stage; hydrofracturing of rocks at depth has not yet been demonstrated as a reliable technology. This is a necessary first step, since the costs of drilling deep holes are high and developers would need to have confidence that a resource of reasonable lifetime could be expected with a high degree of probability before embarking on drilling. The resource of heat at depth is truly 'renewable' only on a very long timescale, since the thermal conductivity of rocks is low and the local temperature will fall as the heat is extracted. The lifetime of a given installation might be a few decades, after which this 'heat mine' would effectively be exhausted and a new installation would have to be drilled on an undisturbed rock site. A typical installation might have an output of 12 MW(e); the station efficiency may be below 10%, so about 90% of the heat would be wasted—which is why the possibility of combined heat and power schemes is being researched (ETSU 1985).

The present programme in the UK involves the continuation of the work in Cornwall at a depth of 2 km, and a decision will be taken at a later date whether or not to proceed to a prototype installation at 6 km. The development is justified by the size of the potential resources, which could be the largest of all the renewable resources, and would, moreover, deliver reliable power at all times, which wind generators would not. Nevertheless, the research is at quite an early stage and the promise may not be realized; both feasibility and costs remain to be demonstrated. A decision whether or not to go to a commercial prototype stage will be taken in 1990 (Department of Energy 1988).

The environmental impact of a 'hot dry rocks' site will mainly result from the chemical nature of the water brought to the surface once the construction and drilling have been completed. This will vary from site to site. Large geothermal plants with an output of 100 MW(e) exist in areas with natural steam sources in the USA, New Zealand, Italy, and Mexico. These plants emit thousands of tons of hydrogen sulphide, ammonia, and carbon dioxide a year, but these effluents are typical of natural sources from volcanic regions and water brought up from wells drilled into stable rock systems should be much cleaner. ETSU (1985) considered that the only significant environmental consequences from a 'hot dry rock' installation for electricity production would be the visual intrusion from the cooling towers, and the need to provide for the cooling water requirements, which may be greater than those for an equivalent conventional generating station.

## Nuclear fusion

The release of energy from nuclear fusion depends on the tighter binding of the nuclei of heavier atoms compared to the nuclei of the lightest ones; energy can be released by inducing the lighter atoms to 'fuse' together. But there is no convenient nuclear reaction that brings this about, like the reaction of neutrons with the heaviest nuclei that induces fission; fusion can only be accomplished by arranging for light nuclei to collide at temperatures typical of those in the centre of stars, about a hundred million degrees. Recent reports of fusion reactions in palladium electrodes at low temperatures have not been substantiated.

Nearly all work has concentrated on the reaction between the two isotopes of hydrogen, deuterium (D or H-2) and tritium (T or H-3):

$$D + T \rightarrow He + n + 17.6 \text{ MeV}$$

Deuterium can be separated from natural hydrogen—there are 33 grams in every $m^3$ of water—but tritium does not exist in nature in significant quantities since it is radioactive with a half-life of 12.3 years. Fortunately, the required tritium can be produced by the reaction of neutrons with either of the isotopes of lithium, Li-6 or Li-7:

$$n + \text{Li-6} \rightarrow \text{He-4} + T + 4.8 \text{ MeV}$$
$$n + \text{Li-7} \rightarrow \text{He-4} + T + n - 2.5 \text{ MeV}$$

The most favoured approach to the problem of producing mixtures of deuterium and tritium at very high temperatures and isolating them for long enough for the reaction between them to take place is to ionize them to form a 'plasma' of charged particles which can then be isolated from the walls of a containing vessel by the application of a strong magnetic field. The neutrons will escape and can be captured by layers of lithium or lithium compounds arranged around the primary vessel, forming the tritium needed to sustain the reaction. A fusion reactor of this type is therefore another type of 'breeder' reactor; the overall reaction of the entire system can be written as:

$$D + \text{Li-6} \rightarrow 2\text{He-4} + 22.4 \text{ MeV}$$

Lithium is fairly plentiful in the earth's crust and supplies should be enough to sustain a large fusion programme. The potential size of the resource and the attractions of a nuclear breeding cycle which does not result in the production of fission products and actinides of long half-life have led to a large international research effort on fusion. Considerable progress has been made towards the realization of experimental conditions under which these reactions can be demonstrated. The JET device at Culham is the leading experiment in an EEC programme on fusion, aimed at establishing the scientific feasibility of attaining fusion by the magnetic confinement of a D–T plasma. If success is achieved, another and larger machine would be needed to establish the technological and engineering feasibility of a true breeder reactor. The economic feasibility of a fusion reactor would require further development beyond that. Thus it is not likely that fusion will be able to make a contribution to energy resources for many decades to come.

Magnetic confinement is not the only way of achieving controlled fusion which is being explored. Another possibility is to heat small quantities of a deuterium–tritium mixture to high temperatures very rapidly by focusing a high-power laser beam, or a beam of charged particles. The pellet targets used are typically glass microballoons of 0.1 mm in diameter, and very high laser powers are necessary. Work on these 'inertial confinement' methods is not as far advanced as that on the magnetic confinement methods.

Detailed studies of possible designs of fusion reactors based on the deuterium–tritium reaction and the lithium breeder cycle have been made in parallel with the experiments. Most of the recent designs have been based on the tokomak system, in which the heating plasma is confined within a steel doughnut structure, and the environmental impact can be estimated, although much development work remains to be done. A fusion reactor with an output of 1 Gw(e), comparable to that from a modern commercial fission reactor, will be a large, complex machine. The reacting plasma would be confined at low pressure within a steel ring of outer diameter 20–30 m and inner diameter 4–10 m, which would be surrounded by the lithium blanket, and coils of super-conducting material to generate the magnetic fields. There would have to be gaseous injection and exhaust systems, a chemical separation plant to extract tritium from the lithium blanket, and a heat extraction system feeding into the steam turbines for electricity generation (Davenport 1983). The whole machine would have to be surrounded by a biological shield, because of the neutron flux and the resulting activation of the materials. A containment building would be necessary because of the possibility of a leak of tritium. The size of the installation and the requirements for land would be similar to that of a large fission reactor; the steam-raising, electricity generation, and low-grade heat rejection circuits would, of course, be exactly the same.

**The environmental impact of fusion: normal operation**

The worst biological hazard that would arise from a fusion reactor would result from the escape of tritium, and the most dangerous form is as tritiated water, HTO, which is more dangerous than the gas, HT. Tritiated hydrogen or water behave in the same way as the normal materials; tritiated water would have a biological half-life of about 10 days in the body (Hancock and Redpath 1985). An ingested dose of 0.01 g of HTO would be enough to lead to severe radiological illness, with a 50 per cent chance of causing early death. Hydrogen diffuses readily through many structural materials and the containment of tritium gas will therefore demand high engineering standards.

The total tritium inventory for a large (1 GW(e) fusion reactor may be of the order of 10 kg, or 3.7 million TBq, including amounts in the reaction zone, the breeder blanket, and the processing plant which purifies the exhaust gas. Daily leakages of the order of 1 part per million, 3.7 TBq, might be acceptable, leading to individual doses to people at the site boundary of no more than 0.01 mSv/y. Such a target may be compared with discharges of tritium actually made from conventional nuclear plants today. Sellafield discharges about 6 TBq per day, and the Canadian CANDU reactors, which have a total tritium inventory of about 2 kg/GW(e), lose about the same amount. The technology to meet these low discharge targets should therefore be available, though demonstration of this can only come from the operation of pilot plants handling large quantities of tritium.

The other radiological consequences occur because of the flux of fast neutrons produced in the D–T reaction. The neutron flux emerging from the reaction zone will be of the same order as the neutron flux in the core of a fast fission reactor, but the neutron energy will be about five times higher. Consequently the material used for the 'first wall' of the plasma containment will be subjected to a very high rate of radiation damage and radioactive species will be formed by neutron activation reactions just as they are in the structural material surrounding the core of a fission reactor. The high damage rate causes severe materials problems and it seems that the structural materials comprising the first wall and blanket will have to be replaced every five years or so (Hancox and Redpath 1985). This operation will involve handling many tens of tons of highly active material and will have to be conducted remotely, and the active material stored on the site until it is removed for disposal or for recycling after a suitable decay period.

The level of radioactivity generated by these activation reactions depends on the materials of construction and on the trace element concentrations in the alloys used. The present-day choice of material is likely to be austenitic stainless steels, and the activation products will be similar to those found in the internal steel structures of fission reactors, including isotopes of iron, cobalt, nickel, and manganese. The total radioactivity in the first wall and blanket, if lithium is used as the breeder material, could be as much as 1 500 000 TBq, which is comparable with the total inventory of a fission reactor. However, the activity would have different characteristics. In the case of a fission reactor, the activity arises mainly from the fission products and actinides in the fuel and therefore from comparatively long-lived species (see p. 158), while the activity in the structural material from a fusion reactor decays more quickly, and should reduce by a factor of more than a million in a hundred years (CEC 1986). The development of new alloys based on vanadium and titanium for this duty could reduce significantly the levels of radioactivity to be handled.

If stainless steels were used for the first wall structure, their initial activity would be as high as spent fuel elements, about 3.7 $TBq/m^3$, and a period of cooling for decades would be required before they could be treated as intermediate level wastes. There would also be low-level wastes arising from the processing systems. Thus the eventual waste disposal facility that would be required would be similar to that required for intermediate and low-level wastes from fission, with the advantage that the activity is comparatively short-lived, so that the proof of safety for the long-term future is simpler.

Decommissioning of a fusion reactor would be much the same type of operation as decommissioning a fission reactor after the fuel itself has been removed. What remain in both cases are large volumes of low-activity material; the quantities of steel and concrete to be dealt with will be similar or perhaps rather larger in the case of fusion (Hancox 1986).

### Safety considerations

The low inventory of fuel in the reaction zone of a fusion reactor and the relatively low power density lead to the conclusion that a large-scale accident with a catastrophic release of radioactivity is not possible. The numbers of the public at risk after the most serious possible accident are likely to be lower than in the case of a fission reactor by factors of ten or a hundred (Häfele 1977), and the contamination of surrounding areas after a large release should be less persistent.

Nevertheless, it will be necessary to show by probabilistic analysis that the probability of accidents leading to a significant release of activity is acceptably small; such an exercise cannot be carried out until the design of a full-scale reactor is far advanced. The main hazard to the public will arise from the large inventory of tritium, and the possibility of breaching one of the reactor circuits containing tritium and the surrounding containment. If liquid lithium is used as the breeder material, there will be a possibility of a fire or of a lithium-water reaction producing hydrogen, but this possibility would be eliminated by the use of solid lithium compounds, with an inert gas such as helium as a heat-conducting fluid. Accidents could be initiated by external events such as aircraft crashes or earthquakes or by internal events such as electro-mechanical failures, misdirection of the plasma, or the breakdown of equipment. The decay heat due to the neutron activation products might amount to 2% of full power 10 minutes after shut-down, and some measures to retain the capacity to cool the most active parts of the structure after a loss of coolant accident would again be required, to avoid any danger of melting. This requirement could be minimized by the development of structural materials with less ability to capture neutrons.

All in all, a fusion reactor appears to have an inherently lower hazard potential than fission reactors, though large quantities of radioactivity are involved and detailed safety studies on actual designs will have to be carried out (Hancox and Redpath 1985).


## Epilogue

The environmental effects of electricity generation will have to be considered as part of the wider debate on international issues in energy policy. Gradual changes in the balance of technologies used in the supply and use of energy in the first half of the next century are inevitable as supplies of hydrocarbon fuels, oil, and gas, decline at a time of growing world population, which will be largely concentrated in countries that have a great incentive to increasing industrialization to improve a low standard of living. What has been remarkable in recent years has been the growing realization that environmental issues must play an important part, and perhaps even a dominant role, in deciding the optimum response to that challenge.

The use of electricity will probably continue to increase in proportion to industrial activity because of its flexibility as a power source and of its efficiency and cleanliness at the point of use. Electricity demand has been increasing in the world in recent years and shows a sharp increase in countries now described as 'Developing' countries which are pursuing a policy of industrialization. From the point of view of environmental impact, the control of gaseous pollution is more easily tackled in a comparatively small number of power stations than in a widely disseminated number of small installations. For example, increasing use of electricity in transport could reduce the pollution resulting from petrol or diesel engines, and burning coal in power stations can be a cleaner operation than burning coal in domestic or in small industrial furnaces.

Increasing reliance on coal in place of hydrocarbons would seem at first sight to be a sensible reaction to the future supply position. The ecological consequences of burning coal in power stations due to sulphur dioxide and nitrogen oxide emissions, can be minimized by use of the technologies described in Chapter 2. But there seems little chance of trapping and disposing of the carbon dioxide produced, and the extent to which coal is used in the future may well be determined by the emphasis given to the possibility of significant and potentially harmful climatic change due to the 'greenhouse effect'. This has now reached the international political agenda. An important conference on *The changing atmosphere* was held in Toronto on 27–30 June 1988, and was attended by government representatives and scientists from forty eight countries as well as from the UN and international bodies. The conference considered the evidence for climate warming and the predictions of associated changes in sea-level and in precipitation patterns arising from the greenhouse effect, as well as the stratospheric ozone depletion problem and urban air pollution. One of the major recommendations was that the wealthy industrialized countries should reduce emissions of carbon dioxide by 20 per cent of the 1988 levels by the year 2005, though even larger reductions might be necessary in later years. About one-half of this reduction would be sought from increased energy efficiency and other conservation measures, while the other half would be achieved by modification of energy supply (Everest 1988).

While considerable uncertainties on the course and particularly on the local consequences of global climatic change will remain for some time, it would clearly be prudent to contemplate plans to minimize dependence on burning coal. Such a strategy would include measures to reduce the use of electricity for heating by the use of combined heat and power schemes, perhaps preferably burning oil or gas while such a course can be justified by guaranteed supplies, and developing those alternative technologies which promise to generate electricity at around the same cost as burning coal, in order to reduce the economic burden of the change in fuel mix as far as possible. Nuclear power is an already developed technology, and others that show early promise in the UK are tidal power schemes and onshore wind generation (pp. 181–2).

Nuclear power is available now and is already making a significant world-wide contribution to the reduction of atmospheric pollution, replacing some 600 000 million tons of coal a year. It is particularly suited to the generation of base-load electricity and to exploitation in those highly industrialized countries which will have to carry the primary responsibility for reducing carbon dioxide emissions. About 16 per cent of the world's electricity is now generated by nuclear means, with over 400 reactors operating in 25 countries, and the proportion of nuclear generation is as high as 70 per cent in some countries such as France. But further use of the nuclear option is inhibited by fear of radiation and of accidents causing widespread contamination, and by doubts about the safety of waste disposal and the proliferation of nuclear weapons.

The diversion of fuel material from civilian to military use is prohibited under the Non-Proliferation Treaty and there is a surveillance mechanism administered by the International Atomic Energy Agency to monitor compliance. All the states known to possess nuclear weapons (USA, USSR, UK, France, and China) obtained them before building a civil nuclear programme, and not as a result of having done so. There are simpler and more direct methods of obtaining the fissile material needed in weapons than diversion from a civil programme which is increasingly based on high burn-up oxide fuels, and the need for international action to monitor the clandestine manufacture of nuclear weapons would remain even if there were no civil nuclear programme.

The effects of radiation, the chance of an accident, and the safety of waste disposal were discussed in Chapter 3. It is clear that nuclear plants are environmentally benign in normal operation, making only a very small alteration to the natural environment. Even a reactor accident as serious as Chernobyl is calculated to cause a total collective dose commitment over many decades to the inhabitants of the northern hemisphere of only 6 per cent of what they receive every year from background radiation (Clarke 1988). However, the local and regional environmental consequences of Chernobyl were serious and new nuclear stations can only be built if the public are convinced that a satisfactory standard of safety will be achieved and that the possibility of such a large-scale accident is extremely low.

The circumstances of the nuclear industry and the public apprehension it engenders has caused rapid development of the technical process of risk assessment and an equal effort on the political and sociological questions raised by the need for fair public scrutiny of the results (Roberts, 1987). Questions such as how can risk assessments be validated, what is an acceptable degree of residual risk, how does the regulatory apparatus operate, how are local authorities consulted, and many more have been discussed and researched in considerable depth. The nuclear debate, as it is called, has raised many fundamental questions concerning the procedures that should be followed in a democratic society to justify the acceptance of any industry with a potential for causing harm (O'Riordan 1987). The most complete public airing of these issues in this country was the protracted public inquiry into the proposal to site a PWR at Sizewell, which lasted for twenty six months, and covered safety and economic issues. The Inspector's report drew attention to many areas where improved procedures might facilitate both communication with the public and the efficiency of the regulatory processes (Layfield 1987). Examination of the Inquiry and the manner in which it was conducted raises other questions concerning the efficacy of the Inquiry process itself as an important arm of planning policy (O'Riordan et al. 1988).

These questions of what planning procedures should be used will become more urgent as the debate on energy policy is broadened to encompass the environmental issues of pollution and of possible climatic change as well as questions of economics and of local safety. Indeed, nuclear safety is already seen as an international matter in the post-Chernobyl era, and the IAEA and the nuclear operators are taking some steps to achieve uniform standards of safety. Sulphur dioxide emissions from power stations are also the subject of international agreements. Energy policy more generally is increasingly becoming an international question as the threat of global climatic change is perceived as significantly probable. The resolution of the differing economic needs of regions and countries against a background of global change will call for international statesmanship of a high order. The conventions recently held in Montreal on the ozone issue and in Toronto on the 'greenhouse gas' effects have served to emphasize the global consequences of changes in the world's atmosphere, though further research and monitoring is urgently required to enable predictions to be made with greater certainty. While priority must be given to measures aimed at increasing the efficiency of use of all forms of energy to minimize the burden on the world's atmosphere, stable and reliable supplies of electricity will remain a vital component of modern life and of the industrial economy. The choice of supply technologies for the future will have to take into account the possible local and global environmental consequences as well as the more conventional economic considerations, and the probability of these consequences will have to be judged on an assessment of the arguments that have been summarized in this book.

# References

Australian Department of Resources and Energy (1985). A review of the photovoltaics sub-programme. Australian Gov. Pub. Services (Canberra).

CEC (1986). Environmental impact and economic prospects of nuclear fusion. Report EURFU BRU/XII-828/86. CEC Brussels.

Clarke, R.H. (1988). UNSCEAR. *Radiological Protection Bulletin* No. 95, p. 7. National Radiation Protection Board, Chilton, Oxon.

Coppola, A. and Hall, R.E. (1981). A risk comparison. NUREG/CR-1916, US Nuclear Regulatory Commission, Washington D.C. 20555.

Davenport, P.H. (1983). *Nuclear Power Technology*, 1, p. 416. Clarendon Press, Oxford.

Department of Energy (1979). Combined heat and electrical power generation in the UK. Energy Paper No. 35. HMSO, London.

Department of Energy (1988). Energy Paper No. 55. HMSO, London.

ETSU, (1985). Report R30 Prospects for the exploitation of the renewable energy technologies in the UK. HMSO, London.

Everest, D.A. (1988). The greenhouse effect: issues for policy makers. Roy. Inst. International Affairs, London.

Forrest, R., Heap, C. and Doggart, J. (1985). Small scale combined heat and power. ETSU Report. Energy Technology Series 4. Harwell, Oxfordshire.

Glover, R.S. (1981). The environmental impact of the proposed severn barrage. NERC Inst. for Marine Environmental Research, Plymouth.

Häfele, W., Holdren, J.P., Kessler, G. and Kulemski, G.L. (1977). Fusion and fast breeder reactors Report RR-77-8, IIASA, Laxenberg, Austria.

Hancox, R. (1986). Technical Workshop on Fusion Reactor Safety. IAEA, Vienna.

Hancox, R and Redpath, W. (1985). Fusion reactors—safety and environmental impact. *Nuclear Energy*, **24**, p. 263.

Huest, J. (1983). *Nuclear Energy*, **22**, p. 251.

Layfield, Sir Frank. (1987). Sizewell B Public Inquiry. Dept. of Energy Report. HMSO, London.

Okrent, D. (1981). *Proc. Roy. Soc. London*, **A376**, p. 133.

O'Riordan, T. (1987). *Nuclear Technology*, p. 257.

O'Riordan, T., Purdue, H.M. and Kemp, R.V. (1988). *Sizewell B: An anatomy of an inquiry*. Macmillan.

Pooley, D. (1987). Prospects for renewable energy sources in the United Kingdom. *Proc. Roy. Soc., Edinburgh*, **92B**, p. 73.

Roberts, L.E.J. (1987). *Nuclear Energy*, **26**, p. 349.

# GLOSSARY OF TERMS

**AGR**  Advanced Gas Cooled Reactor

**ALARA**  the need to keep radiation exposures *As Low As Reasonably Achievable* — one of the ICRP principles of radiological protection.

**actinides**  a series of heavy chemical elements, all of which are radioactive. It includes the elements uranium and plutonium.

**ACTRAM**  Advisory Committee on the Transport of Radioactive Materials (UK).

**alpha particles ($\alpha$)**  see radiation.

**BEIR**  Committee on the Biological Effects of Ionizing Radiation, National Academy of Sciences, USA.

**beta particles ($\beta$)**  see radiation.

**BNFL**  British Nuclear Fuels plc.

**BWR**  boiling water reactor, a type of nuclear reactor.

**CCN**  cloud circulation nuclei, the seeds on to which water condenses to form droplets in clouds.

**CEGB**  Central Electricity Generating Board (England and Wales). The other generating boards in the UK are the South of Scotland Electricity Board, the North of Scotland Hydro-Electric Board and the Northern Ireland Electricity Service.

**cladding**  the can in which nuclear fuel is sealed to protect it and retain its fission products.

**CO$_2$**  carbon dioxide.

**collective dose**  see radiation dose.

**control rods**  rods of a neutron-absorbing substance which are used to control the reactivity of a reactor. The degree to which the control rods are inserted into the core of the reactor controls the number of free neutrons available and hence the rate at which the chain reaction proceeds.

**core**  the assembly of fuel in a nuclear reactor.

**critical group**  a term used in radiological protection for those members of the public likely to be exposed to the largest dose of radiation as a result of any particular operation.

**decay heat**  heat produced by the radioactive disintegration of fission products in nuclear fuel, which continues even after the fission reaction has stopped.

**decay series**  a series of elements, each member of which is formed by the radioactive decay of the previous member.

**design basis event**  the maximum expected event (e.g. maximum load) that a structure is designed to resist with an adequate safety margin.

**dose**  see radiation dose.

**DMS**  dimethyl sulphide.

**EPRI**  Electrical Power Research Institute, USA.

**ERL**  Emergency Reference Level: the level of radiation dose at which public authorities are required to consider protective measures in the event of an accident.

**fast reactor**  a reactor in which the chain reaction is sustained by fast neutrons. In fast breeder reactors, new fuel is bred by the absorption of surplus neutrons in the non-fissile isotope U-238.

**FGD**  fuel gas desulphurization at power stations.

**fission**  the spontaneous or induced disintegration of the nucleus of a heavy atom into two or more lighter ones (the fission products), with the release of energy.

**fissile**  isotopes which undergo the fission reaction, for example U-235, Pu-239.

**fuel reprocessing**  the process whereby spent nuclear fuel is separated into uranium, plutonium, and radioactive waste products.

**fusion**  the uniting of two light nuclei into one heavier one, with the release of energy.

**fuel flask**  transport container used for transporting nuclear fuel.

**gamma radiation ($\gamma$)**  see radiation.

**GCM**  general circulation model of the world's atmosphere.

**Gigawatt (GW)**  a unit of power, one thousand million watts.

**(GW(e)**  a gigawatt of electrical power.

**Gt: gigatonne**  one thousand million metric tons ($10^{15}$ grams).

**HSE**  Health and Safety Executive (UK).

**IAEA**  International Atomic Energy Agency, based in Vienna, consisting of 111 countries.

**ICRP**  International Commission on Radiological Protection.

**INFCE**                International Fuel Cycle Evaluation, an international study which reported in 1980.

**ionization**           see radiation.

**isotopes**             nuclei of a given chemical element with different atomic masses, resulting from the presence of different numbers of neutrons.

**LPG**                  liquefied petroleum gas.

**magnox**               a magnesium alloy used as fuel cladding in some gas-cooled reactors, hence 'magnox reactor'.

**NEA/OECD**             Nuclear Energy Agency of the Organization of Economic Co-operation and Development.

**neutrons**             see radiation.

**NII**                  Nuclear Installations Inspectorate, part of the Health and Safety Executive, responsible for the licensing of UK nuclear plant.

**NO, NO$_2$**           gases which are oxides of nitrogen, also referred to as 'NO$_x$'.

**NPT**                  Non-Proliferation Treaty.

**NRC**                  Nuclear Regulatory Commission, USA.

**NRPB**                 National Radiological Protection Board, UK.

**nuclear reactor**      power-producing unit using nuclear energy.

**probabilistic risk assessment/ probabilistic safety assessment (PRA/PSA)**    a mathematical technique used to determine the probability of faults occurring and their possible outcome.

**ppb**                  parts per billion, i.e., parts per thousand million.

**ppm**                  parts per million.

**pH**                   measure of acidity on a scale from 1 (acid) through 7 (neutral) to 14 (alkaline).

**PWR**                  nuclear reactor cooled by water under pressure.

**radiation**            the term used in this book for ionizing radiation, that is, particular or electromagnetic radiation that produces ionization (the removal of one or more electrons from an atom, leaving positively charged ions) in material through which it passes. The ionizing radiations are:

    alpha particles: the nuclei of atoms of the element helium;
    beta particles: electrons;
    gamma radiation and X-rays: electromagnetic radiations similar to light and radio waves;
    neutrons: neutral particles present in all nuclei except hydrogen.

| | |
|---|---|
| **radiation background** | the natural radiation to which people are exposed. |
| **radiation dose** | the quantity of radiation absorbed by an individual or by a specific organ. Collective dose is a measure of the radiation dose received by a population; it is the sum of the doses received by all the individuals in that population. |
| **radiation dose equivalent** | a measure of the biological effectiveness of a dose of radiation. |
| **radiological protection** | the protection of individuals from the effects of ionizing radiation. |
| **RBMK** | a graphite-moderated water-cooled reactor (especially in USSR). |
| **RWMAC** | Radioactive Waste Management Advisory Committee, UK. |
| **SO$_2$** | sulphur dioxide (gas). |
| **somatic** | relating to the body of an animal as distinct from the germ cells. A somatic effect of radiation is one affecting the individual receiving the radiation, as distinct from a hereditary effect, which affects subsequent generations. |
| **specific activity** | the radioactivity per unit mass or volume. |
| **spent fuel** | nuclear fuel with a lowered fissile content after irradiation in a reactor. |
| **stochastic** | governed by statistics. The stochastic effects of radiation are those in which the dose governs the probability of the effect occurring, not the severity of the effect itself. The non-stochastic effects are those in which the dose governs the severity of the effect. |
| **thermal reactor** | a reactor in which the chain reaction is sustained by thermal neutrons, neutrons which have been slowed down by passage through a moderator. |
| **UNSCEAR** | United Nations Scientific Committee on the Effects of Atomic Radiation. |
| **UKAEA** | United Kingdom Atomic Energy Authority. |
| **USNRC** | US Nuclear Regulatory Commission. |

# INDEX